SpringerBriefs in Law

More information about this series at http://www.springer.com/series/10164

Markus Adam

Accelerating E-Mobility in Germany

A Case for Regulation

Springer

Markus Adam
Hamburg
Germany

ISSN 2192-855X ISSN 2192-8568 (electronic)
SpringerBriefs in Law
ISBN 978-3-319-44883-1 ISBN 978-3-319-44884-8 (eBook)
DOI 10.1007/978-3-319-44884-8

Library of Congress Control Number: 2016948240

Printed on acid-free paper

This Springer imprint is published by Springer Nature
The registered company is Springer International Publishing AG
The registered company address is: Gewerbestrasse 11, 6330 Cham, Switzerland

To Sandra

Acknowledgement

First of all, I would like to thank everyone who supported me in writing this paper.

Gratitude is owed to Prof. Dr. Jörg Philipp Terhechte and Prof. Dr. Thomas Wein, Professors at the Leuphana University of Lüneburg, who inspired and fostered this paper.

Thanks are due to my friends Anja, Christina, Vera, Alexander and Carl-Friedrich for the numerous talks and the important feedback on this study. I would like to thank my colleagues at LichtBlick for many discussions and scrutinising this work. I owe a great debt of gratitude to Gero Lücking, my boss at LichtBlick, who supported my ideas right from the beginning and who made it possible for me to present the ideas of this paper to the German Energy Industry on many occasions. I am very grateful to Vivien and Antony for giving feedback and proofreading my work.

Furthermore I would like to thank Springer and Dr. Brigitte Reschke for including my work in the *SpringerBriefs in Law* series and the very efficient realisation of this publication.

I am thankful to my parents, my sister and my brother-in-law for the encouragement and the support from afar.

My very special thanks go to my wonderful wife Sandra, without whom I could not have managed writing this paper. Your loving and everlasting support made this publication possible in the first place. I could not have done it without you.

Contents

Abbreviations

BDEW	German Association of Energy and Water Industries
BEV	Battery Electric Vehicle
CCS	Combined Charging System
CFI	Court of First Instance
ECJ	European Court of Justice
EmoG	Electromobility Act (Elektromobilitätsgesetz)
EnWG	Energy Industry Act (Energiewirtschaftsgesetz)
EStG	Income Tax Act (Einkommensteuergesetz)
EVSE	Electric Vehicle Supply Equipment
FCJ	Federal Court of Justice
GWB	Act against Restraints of Competition
KAV	Ordinance regarding concession fees (Konzessionsabgabenverordnung)
KraftStG	Law on Car Tax (Kraftfahrzeugsteuergesetz)
NAPE	National Action Plan Energy Efficiency
NEP	National Development Plan Electromibilty
NPE	National Platform Electromobility
StromNZV	Ordinance Regulating Network Access (Stromnetzzugangsverordnung)
VKU	Association of Municipal Utilities (Verband der kommunalen Unternehmen)
VwVfG	Administrative Procedure Act (Verwaltungsverfahrensgesetz)
WACC	Weighted Average Cost of Capital

List of Figures

List of Tables

Chapter 1
Electromobility—The Current State

1.1 Introduction

E-Mobility, short for Electromobility, means the electrification of individual transport. Unlike mass transport, which was electrified long ago in the case of trains and cable cars, individual transport, i.e. cars, trucks and motorcycles, has been dominated by the combustion engine for over one hundred years. But meanwhile a number of manufacturers have emerged that offer electric vehicles from series production. According to data of the Federal Motor Transport Authority 6,051 new electric vehicles were registered in 2013.[1] The number of first time registrations rose by 40.8 % to 8,522 electric vehicles in 2014.[2] The Federal Motor Transport Authority defines electric vehicles as vehicles with a solely electric powertrain, although no distinction is made between a pure battery electric vehicle (BEV) and a BEV with a range extender, which is a combustion engine for charging the battery.[3] In the year 2014 BEVs accounted for a vanishingly low 0.28 % of the total new passenger car registrations in Germany.[4] Including the month of September 7,385 new BEVs have already been registered in 2015, an increase of 22.1 % compared to the same period of the previous year.[5]

[1]Federal Motor Transport Authority (2014).
[2]Federal Motor Transport Authority (2015c).
[3]Federal Motor Transport Authority (2015a, p. 3).
[4]Federal Motor Transport Authority (2014).
[5]Federal Motor Transport Authority (2015b, p. 12).

M. Adam, *Accelerating E-Mobility in Germany*, SpringerBriefs in Law
DOI 10.1007/978-3-319-44884-8_1

1.2 Legal Framework

The following short overview of the legal framework shall highlight the legal foundation for E-Mobility, which is already in place and which suggests that E-Mobility will probably not be a temporary phenomenon.

1.2.1 National Measures

As early as 2009 the federal government adopted a "National Development Plan Electromobility" (NEP) in which the central objective of one million electric vehicles on Germany's roads in 2020 was declared.[6] The federal government reaffirmed this goal in 2012[7] and reconfirmed in the coalition agreement of 2013 the call for establishing a lead market for electric mobility.[8] Deciding on the NEP in 2009 is remarkable in so far as that electromobility still led an absolute niche existence at that time: According to figures from the Federal Motor Transport Agency by the end of 2009, only 1,588 electric vehicles in Germany had been registered.[9]

The NEP proclaims 19 concrete goals, of which the first four are to be quoted here because of their relevance to the issues discussed herein:

1. *"Electric mobility shall make a significant contribution to achieving climate protection targets.*
2. *Through the use of renewable sources to meet the energy demands of electric vehicles a contribution can be made to the implementation of the expansion targets for renewable energies and for improved integration of fluctuating energy producers into the grid. This contributes to an increasing security of supply in the long term.*
3. *The power grids in Germany should become more efficient through the use of modern information technologies and the integration of electric vehicles.*
4. *The additional demand for electric energy in this sector is to be covered by electricity from renewable sources. For that, the otherwise unusable power from fluctuating renewable energies should be used primarily as part of the load management by electromobility. Further expansion potentials of renewable energies need to be developed for going beyond current demand for electric mobility."*[10]

Thus, the NEP formulates within its first goals the energy industry relevance of this new transport sector and addresses the key challenges for the energy sector.

---

[6]Federal Ministry of Education and Research (2009, p. 2).

[7]The Press and Information Office of the Federal Government (2012).

[8]CDU/CSU Parliamentary Group in the German Bundestag (2013, p. 15).

[9]Federal Motor Transport Authority (2010, p. 24).

[10]Federal Ministry of Education and Research (2009, pp. 17–18).

Moreover, the government program for 2011 created so-called "showcase regions", i.e. state-subsidized regions where e-mobility projects with various focus areas were specifically implemented. The "National Platform for Electromobility" (NPE) acts as an advisor on electric mobility for the federal government. The NPE is a committee consisting of representatives from industry, science, politics, trade unions and associations. There are seven working groups in the NPE, which analyse key topics and make recommendations to the government.[11]

A first legislative initiative towards e-mobility took place in 2012, when an exemption of electric vehicles from motor vehicle tax was decided upon, see § 3d KraftStG. In 2013 the Law regarding company car taxation was amended, which led to a lesser tax load for employees driving an electric company car, see § 6 Abs. 1 Nr. 4 EStG.[12]

The most recent legislative initiative, the Electromobility Act (EmoG), allows for special facilities in road transport for e-vehicles, e.g. municipalities get the possibility to grant special parking rights or the use of bus lanes to e-vehicles, see § 3 EmoG.

Not least, the Programme of Action for Climate Protection of the Federal Ministry for the Environment, Nature Conservation, Building and Nuclear Safety stresses the great importance of e-mobility for reaching climate targets in Germany.[13] The action program describes the planned measures by Germany in order to achieve the national target to reduce greenhouse gas emissions by 2020 by at least 40 % compared to 1990.[14] In regard to the CO_2 reduction targets it is stated:

> *"The consistent introduction of electric mobility in passenger transport—with electricity from renewable sources—in particular with regard to the medium and long-term climate change targets is of very great importance."*[15]

As a first, concrete step to promote e-mobility a special amortisation for companies and tradesmen when purchasing an electric vehicle is planned.[16] This would be the first direct subsidy to a great extent for this market.

All in all it can be said that the federal government despite all the criticism, e.g. in regard to the lack of direct support for private purchase of e-vehicles, adopted the topic of e-mobility very early and in many fields.

---

[11]Federal Ministry for the Environment, Nature Conservation, Building and Nuclear Safety (2010).

[12]The list price of the e-vehicle is reduced by 450 EUR/kWh of the battery capacity (to a maximum of 9.500 EUR) and the reduced price is basis for the 1 %-rule of the company car taxation. Therefore the tax burden for the employee is lowered significantly.

[13]Federal Ministry for the Environment, Nature Conservation, Building and Nuclear Safety (2014).

[14]Federal Ministry for the Environment, Nature Conservation, Building and Nuclear Safety (2014, p. 6).

[15]Federal Ministry for the Environment, Nature Conservation, Building and Nuclear Safety (2014, p. 22).

[16]Federal Ministry for the Environment, Nature Conservation, Building and Nuclear Safety (2014, p. 56).

1.2.2 *European Measures*

The European Union adopted the climate and energy package in 2009,[17] which set the following three key targets to be met by the EU in the year 2020:

- a reduction of greenhouse gas emissions in the EU by 20 % in comparison to the year 1990;
- a share of renewable energies of 20 % of the energy consumption in the EU;
- and a 20 % improvement in energy efficiency, i.e. a 20 % cut in primary energy consumption.

Accordingly, these goals were briefly summarized as “20–20–20 by 2020”. The climate and energy package consists of several legislative proposals on these three topics that are briefly summarized below, as it leads to very concrete points for the promotion of electromobility by the European legislation. These points should also have a direct effect in the next few years due to the not so distant target year 2020.

1.2.2.1 Reduction of Greenhouse Gas Emissions

Already in the Regulation (EC) No. 443/2009 of the European Parliament and of the Council of 23 April 2009 concrete specifications were regulated, which should lead to a reduction of CO_2 emissions from passenger cars and light commercial vehicles. The recitals of the Regulation (EC) No. 443/2009 explicitly reference the goal of reducing CO_2 emissions in the framework of the “Strategy 2020”. The provisions in this regulation were amended in 2014 by Regulation (EU) No. 333/2014.

This regulation package designates a maximum limit of 95 g CO_2/km for CO_2 emissions from passenger cars as a target for the year 2020, see Art. 1 pt. 1 of the Regulation (EU) no. 333/2014. The regulation specifies in Art. 1 pt. 5 a complex calculation mechanism which allows the determination of the average CO_2 emissions of each manufacturer. Vehicles that have CO_2 emissions of less than 50 g/km, count as 2 vehicles and therefore will lower the emission count of the manufacturer even more. A glance at the current values of vehicles with internal combustion engines shows the incentive effect for manufacturers: A VW Golf, equipped with the most fuel-efficient engine, still causes CO_2 emissions of 92 g/km,[18] while even the Toyota Prius with a hybrid engine still shows CO_2 emissions of 70 g/km.[19] The plug-in version of the Toyota Prius shows, however, the positive effect of the larger battery: The CO_2 emissions of the plug-in model are measured at 49 g/km.[20] Therefore, in order to bring a vehicle with an output of less than 50 g CO_2/km on

[17]European Commission (2009).

[18]Volkswagen AG (2016b, p. 29).

[19]Toyota Deutschland GmbH (2016a).

[20]Toyota Deutschland GmbH (2016b).

the market, manufacturers will need to switch to plug-in hybrids and pure electric vehicles. How ambitious the target of 95 g CO_2/km will be for some manufacturers, is shown in the latest figures: According to the figures by the European Environment Agency, the average CO_2 emissions for new cars in 2013 remained below the threshold of 130 g CO_2/km and thus stayed below the target for 2015,[21] but no vehicle with a petrol or diesel engine with CO_2 emissions of less than 95 g CO_2/km—the target number for 2020—is currently available.

In order for the manufacturers to reach the target of 95 g CO_2/km for the new car fleet, a significant change of direction towards alternative drive trains, such as hybrid vehicles and especially electric vehicles, will be needed. The impact of this regulation is already visible: Almost all car manufacturers with a European brand have electric vehicles from mass production on offer for the first time.[22]

Although the amendment of the regulation was actually a softening of the goals of the original version in 2009, the new regulation still has a strong incentive effect and ambitious CO_2 reduction targets and should therefore significantly contribute to the success of e-mobility in the European Union.

1.2.2.2 Promotion of the Use of Energy from Renewable Sources

The conditions for the development of renewable energies in the European Union are governed by Directive 2009/28/EU. There are two key targets named in recital 20 of the Directive: Firstly, the share of renewables of the final energy consumption in the European Union shall account for 20 % in 2020 and secondly the share of renewables in final energy consumption in the transport sector shall amount to 10 % according to the 20–20–20 rule.

Accordingly, the Directive recognizes the importance of renewable energy for the transport sector in other recitals. The goal of a 10 % share of renewables in final energy consumption in the transport sector is intended to be achieved not only with biofuels, but also through new technologies, thereby using diversified renewables.[23] The issue of storing energy from renewable sources is addressed in the recitals of the directive as well and is designated as a topic worthy of support.[24]

The Directive addresses the issue of "e-mobility" in specific support schemes, especially in Art. 3 para. 4, which states that the proportion of renewable energy, which is consumed in road vehicles with an electric drive train, is weighted by a factor of 2.5 for the calculation of the proportion of energy from renewable sources in gross final energy consumption in 2020. Hereby, a significant improvement is made reaching the target of the Directive and it is a considerable incentive for the

---

[21]European Environment Agency (2014).

[22]A good overview of the current offer of e-vehicles in Germany can be found at: http://www.goingelectric.de/elektroautos/.

[23]See recitals 2, 18 und 29 of Directive 2009/28/EU.

[24]See recital 57 of Directive 2009/28/EU.

expansion of electromobility. This betterment of e-mobility is useful because electric vehicles have significantly better energy efficiency than conventional vehicles with combustion engines. Not only can CO_2 emissions be prevented this way, but primary energy can be saved as well.

The privileged grid access for renewable energy is governed by Article 16 of the directive and is a centerpiece of the European legal standards, since the priority (or guaranteed) grid access for renewables ensures the actual promotion of renewables by way of third party access and thus is an important tool for the conversion to a sustainable energy supply in the European Community. Without this network access regulation, the delivery of energy from renewable sources to final customers would be put into serious question; therefore those provisions are an important jump start for renewables in the energy market. This regulatory approach at European and national level will play an important role in regard to the following research of the charging infrastructure for electric vehicles.

1.2.2.3 Improving Energy Efficiency

The topic of "energy efficiency", the third cornerstone of the climate and energy package, is governed by the Energy Efficiency Plan 2011 and the Energy Efficiency Directive, the Directive 2012/27/EU. The Energy Efficiency Plan 2011 stresses the importance of the transport sector in paragraph 7 to the effect that the transport sector is in terms of energy use the fastest growing sector with the greatest dependence on fossil fuels. However, the energy efficiency roadmap dispensed with further clarifications on the measures to be taken and instead refers to the "White Paper on transport."[25]

The White Paper, however, has it all, because as the first target among the 10 targets listed the following is stated:

> „*Halve the use of 'conventionally-fuelled' cars in urban transport by 2030; phase them out in cities by 2050; achieve essentially CO_2-free city logistics in major urban centres by 2030*"[26]

Among the numerous initiatives, which are annexed to the White Paper, a new legal framework for CO_2 requirements in road transport is envisaged (see Sect. 1.2.2.1) as well as a legal framework for the interoperability of charging infrastructure for clean vehicles.

The issue of interoperability of charging structures, i.e. the question what connectors should be used, is regulated by the directive on the on the deployment of alternative fuels infrastructure, Directive 2014/94/EU ("AFI Directive"), which entered into force on November 17, 2014. Pursuant to Annex B of the directive,

[25]European Commission (2011).

[26]European Commission (2011, para. 2.5).

recharging points for electric vehicles shall be equipped with connectors of Type 2 as described in standard EN 62196-2.[27]

In addition, the AFI Directive also contains numerous other provisions in the field of e-mobility, which should allow for significant development. This includes in particular the obligation of Member States to draw up a national strategic framework, which in turn, for example, must provide for the establishment of an adequate number of public charging points. This Directive therefore undertakes the hitherto clearest commitment of the European Union to e-mobility.

Another cornerstone of the "20–20–20 by 2020" program is the Energy Efficiency Directive, Directive 2012/27/EU, which entered into force on December 4, 2012. The aim of this Directive is to save primary energy, i.e. promote a more efficient use of energy. Pursuant to Art. 3, para. 1a) of the Directive the Union's energy consumption must not exceed 1,474 Mtoe of primary energy or must not amount to more than 1,078 Mtoe of final energy in 2020. By comparison, the consumption of primary energy in the Union accounted for 1,666 Mtoe in 2013.[28] Therefore a reduction of around 11.5 % is still required to reach the target for the year 2020.

The Directive obliges Member States to establish national energy efficiency action plans, in which the relevant actions and contributions to achieve these targets are specified.

Since the energy conversion in electric vehicles is done very efficiently with an engine efficiency of 65 %,[29] this technology could significantly help to achieve the goals of energy efficiency in the transport sector. By comparison, the overall efficiency of gasoline and diesel engines is just 15–20 %.[30] Accordingly, incentive measures for electromobility in the form of financial subsidies for the construction of charging stations and the purchase of electric vehicles are included in the National Energy Efficiency Action Plan for Great Britain.[31] The German NAPE in comparison is far more reluctant with direct financial support measures, but a promotion of commercial electric vehicles is at least considered in the form of a special allowance for depreciation.[32]

Moreover, the Directive especially addresses the energy sector and specifies a number of measures in the recitals to ensure the achievement of efficiency targets. In recitals no. 44 and 45, two areas are mentioned with the topics "load control" and "smart grids" in which e-mobility can be an important component. The batteries of electric vehicles can be integrated via an intelligent control system in the electricity supply system and thus allow for more efficient management of the network. In this

---

[27]The connector of Type 2 has already been adopted by German car manufacturers as the standard application.

[28]Eurostat (2015).

[29]Ifeu—Institut für Energie und Umweltforschung Heidelberg GmbH (2009, p. 3).

[30]Golloch (2005, p. 1).

[31]Department of Energy & Climate Change in the UK (2014, p. 38).

[32]Federal Ministry for Economic Affairs and Energy (2014, p. 41).

use case e-vehicles are connected bidirectionally to the grid and may purchase energy from the grid or feed energy back to the grid, depending on the load situation (so-called "vehicle-to-grid technology").

The specific rules are laid down in Art. 15 of Directive 2012/27/EU, which states, inter alia, non-discriminatory market access by third parties to the markets of ancillary services. Providers of demand response solutions, especially aggregators, are entitled for such a claim against the operators of the transmission and distribution grids.

Therefore this Directive could also provide some significant impetus for the breakthrough of e-mobility.

1.3 Advantages of E-Vehicles

At this point, it shall be briefly shown why electric vehicles probably are not a temporary phenomenon, and not only due to the previously mentioned national and European political decision-making, but will rather replace conventional vehicles with combustion engines in the medium to long term in all likelihood:

- E-vehicles are almost silent in city traffic: The electric motors cause almost no noise, compared to gasoline and diesel engines, which are limited to a driving noise of a maximum of 74 (dB).[33] In order to alert pedestrians to the presence of silent electric vehicles, even an obligation to install an Acoustic Vehicle Alert System (AVAS) has been introduced at EU level.[34] But not only the engine noise is omitted; typically electric vehicles also have no gearbox, so that no acceleration noise exists (except for airstream noises). This benefit should not be underestimated for vehicle passengers, road users or local residents. The positive impact on urban development and life in the city will probably be enormous in the long term. Busy streets with heavy traffic will clearly cause fewer emissions and thus bring a significant improvement in living and housing standards for residents.
- E-vehicles have a considerable acceleration potential, since electric motors provide power without any perceptible delay. Even electric vehicles from the small car and midsize car segment have acceleration values, which are otherwise only found in the sports car segment. As a gearbox is unnecessary, there is no gear shift in the classical sense, but rather the settings forward, reverse and neutral, comparable to an automatic transmission. As the author himself was able to find out in test drives with a Tesla Model S and a BMW i3, the

[33]See subparagraph 2.1.1 of Annex A of Directive 2007/34/EC.

[34]See Art. 8 of Regulation (EU) No. 540/2014 of the European Parliament and of the Council of 16 April 2014 on the sound level of motor vehicles and of replacement silencing systems.

acceleration, combined with the silence of driving, leads to superior handling and a better driving experience.[35]

- Compared to conventional vehicles, e-vehicles are built with significantly fewer moving parts, so that electric vehicles are virtually maintenance free. Brake pads must be changed significantly less frequently due to the recuperation. Gearbox, clutch and exhaust systems do not exist, and neither do oil changes. In addition there are lower costs for electric charging compared to refueling with gasoline, diesel or natural gas. The running costs of electric vehicles are therefore significantly lower than for conventional vehicles with combustion engines.[36]
- The energy density of the battery is considerably lower, compared to a gasoline, diesel or natural gas tank. From a purely statistical point of view electric vehicles are therefore likely to be less susceptible to explosive or fire accidents.[37]
- A further advantage is the significantly higher energy efficiency, since most of the energy actually gets to the drive train, unlike with an internal combustion engine, in which a large part of the energy is converted into heat loss.[38]
- But perhaps the strongest argument in the end is the ability to power electric vehicles with green electricity, i.e. to charge with CO_2-free generated electricity, and therefore enable sustainable road transport with virtually no CO_2 emissions.
- Another effect, in addition to reducing CO_2 emissions, would be the reduction of economic dependence on oil imports.[39]
- And last but not least electric vehicles can be integrated intelligently by means of "vehicle-to-grid" applications in distribution networks, i.e. they can provide added value for the distribution network by means of new energy business models for grid stability and the inclusion of fluctuating renewable energies.

1.4 Problem Description

Besides the described benefits of electric vehicles there is, in addition to the currently much higher selling prices, another negative factor and that is the significantly shorter range compared to vehicles with combustion engines. The currently highest range of an electric vehicle series is offered by the Model S of the US car maker Tesla Motors with up to 528 km range per charge through the 85 kWh version with all-wheel drive.[40] However, the typical range of an electric vehicle,

---

[35] See Auto, Motor und Sport (2014).

[36] Auto, Motor und Sport (2012).

[37] According to calculations by Tesla Motors, Inc. (2013).

[38] Ökoinstitut e.V. (2011, p. 9).

[39] Federal Ministry of Education and Research (2009, p. 8).

[40] Tesla Motors, Inc. (2016).

such as the BMW i3[41] or VW e-Golf is closer to 100–200 km.[42] As with conventional vehicles, the driving style and the weather play a significant role for the actual achievable range. Cold weather and faster driving can cause the actual possible range to decrease by 30 % or more compared to the manufacturer's official data.[43] As a result, this means that electric vehicles have a significantly shorter range than conventional vehicles with internal combustion engines. In addition to a higher purchase price, the so-called range anxiety is thus probably the most common counter-argument for switching to an electric vehicle. To enable longer overland trips, a comprehensive and functional charging infrastructure is especially needed (besides larger batteries).

According to the German Association of Energy and Water Industries (BDEW) 5,500 publicly accessible charging points were available in Germany in December 2014.[44] But in addition to the construction and operation of charging stations, efficient access to these charging stations is also needed to offer final consumers a real possibility of charging. Drivers of electric vehicles are provided technical access to almost every public charging station through the legally required standardization of the connectors[45] and in case of doubt an adapter will enable the driver to connect the vehicle to the charging station.

The driver of an electric vehicle, however, is currently faced with four key problems in regard to charging his vehicle on the road:

1.4.1 Conclusion of Contract with the Charging Station Operator

Before the e-vehicle driver can take advantage of the respective charging station, he has to enter, usually in advance, into a contract with the respective charging station operators. With more than 100 charging station operators[46] in Germany, this quickly turns into an obstacle to the charging process.

[41]BMW AG (2016).

[42]Volkswagen AG (2016b, p. 8).

[43]auto, motor und sport (2014).

[44]BDEW (2015a).

[45]See Art. 4 para. 4 of directive 2014/94/EU.

[46]At BDEW alone over 100 charging station operators (EVSE Operator) are registered: https://bdew-codes.de/Codenumbers/EMobilityId/OperatorIdList.

1.4.2 No Free Choice of Energy Supplier

Unlike with the conventional power supply, the driver of an electric vehicle has no way to choose his electricity supplier at a charging station. Accordingly, competition at the final consumer level is at risk.

1.4.3 Excessive Pricing

Prices for charging significantly exceed the comparable average price of electricity for households and commercial customers at the low voltage level. Thus is one of the benefits of e-mobility—the much cheaper driving energy compared to gasoline or diesel fuel—at risk of being lost.

1.4.4 Lack of Charging Infrastructure

There is a lack of adequate charging points for commuters and especially for cross-country trips. Due to the low range of electric vehicles charging stations are urgently needed to enable longer journeys.

1.4.5 Approach to a Solution

These four issues will be discussed from the point of view of end users hereinafter in regard to energy industry, antitrust and economic aspects.

Four theories are formulated and examined in view of the applicable legal framework and discussed so as to identify solutions that would result in significantly improved conditions for charging at public charging points.

References

Auto, Motor und Sport (2012) Elektroautos punkten in der Werkstatt. http://www.auto-motor-und-sport.de/news/studie-zu-wartungskosten-elektroautos-punkten-in-der-werkstatt-6146082.html. Accessed 06 July 2016

Auto, Motor und Sport (2014) E-Autos im Härtetest. http://www.auto-motor-und-sport.de/vergleichstest/elektroautos-6-modelle-im-haertetest-8498096.html. Accessed 06 July 2016

BDEW (2015) Ausbau der Ladeinfrastruktur für Elektroautos in Deutschland auf gutem Weg. https://www.bdew.de/internet.nsf/id/20150410-pi-ausbau-der-ladeinfrastruktur-fuer-elektroautos-in-deutschland-auf-gutem-weg-de. Accessed 06 July 2016

BDEW (2015) EVSE operator list. https://bdew-codes.de/Codenumbers/EMobilityId/OperatorIdList. Accessed 06 July 2016
BMW AG (2016) Reichweite und Laden des BMW i3. http://www.bmw.de/de/neufahrzeuge/bmw-i/i3/2015/reichweite-laden.html. Accessed 06 July 2016
CDU/CSU Parliamentary Group in the German Bundestag (2013) Coalition agreement between the CDU, CSU and SPD. https://www.cdu.de/sites/default/files/media/dokumente/koalitionsvertrag.pdf. Accessed 06 July 2016
Department of Energy & Climate Change in the UK (2014) UK national energy efficiency action plan. https://ec.europa.eu/energy/sites/ener/files/documents/2014_neeap_united-kingdom.pdf. Accessed 06 July 2016
European Commission (2009) Commission welcomes adoption of climate and energy package, press release April 23, 2009. http://europa.eu/rapid/press-release_IP-09-628_en.htm. Accessed 06 July 2016
European Commission (2011) WHITE PAPER—roadmap to a single European transport area—towards a competitive and resource efficient transport system, COM/2011/0144 final. http://eur-lex.europa.eu/legal-content/EN/TXT/HTML/?uri=CELEX:52011DC0144&from=EN. Accessed 06 July 2016
European Environment Agency (2014) Car and van makers continue to meet CO_2 emission targets ahead of schedule, press release of October 30, 2014. http://www.eea.europa.eu/highlights/car-and-van-makers-continue. Accessed 06 July 2016
Eurostat (2015) Consumption of energy—Main statistical findings. http://ec.europa.eu/eurostat/statistics-explained/index.php/Consumption_of_energy. Accessed 06 July 2016
Federal Ministry for Economic Affairs and Energy (2014) Nationaler Aktionsplan Energieeffizienz. http://www.bmwi.de/DE/Mediathek/publikationen,did=672756.html. Accessed 06 July 2016
Federal Ministry of Education and Research (2009) Nationaler Entwicklungsplan Elektromobilität der Bundesregierung. https://www.bmbf.de/files/nationaler_entwicklungsplan_elektromobilitaet.pdf. Accessed 06 July 2016
Federal Ministry for the Environment, Nature Conservation, Building and Nuclear Safety (2010) Nationale plattform Elektromobilität. http://www.bmub.bund.de/themen/luft-laerm-verkehr/verkehr/elektromobilitaet/nationale-plattform-elektromobilitaet/. Accessed 06 July 2016
Federal Ministry for the Environment, Nature Conservation, Building and Nuclear Safety (2014) Aktionsprogramm Klimaschutz 2020. Cabinet decision on December 03, 2014. http://www.bmub.bund.de/fileadmin/Daten_BMU/Download_PDF/Aktionsprogramm_Klimaschutz/aktionsprogramm_klimaschutz_2020_broschuere_bf.pdf. Accessed 06 July 2016
Federal Motor Transport Authority (2010) Bestand an Kraftfahrzeugen nach Emissionen und Kraftstoffen 1. Januar 2010. http://www.kba.de/SharedDocs/Publikationen/DE/Statistik/Fahrzeuge/FZ/2010/fz13_2010_pdf.pdf?__blob=publicationFile&v=6. Accessed 06 July 2016
Federal Motor Transport Authority (2014) Pressemitteilung Nr. 02/2014. http://www.kba.de/DE/Presse/Pressemitteilungen/2011_2015/2014/Fahrzeugzulassungen/pm02_2014_n_12_13_pm_komplett.html?nn=1112104. Accessed 06 July 2016
Federal Motor Transport Authority (2015a) Methodische Erläuterungen zu Statistiken zu Fahrzeug zulassungen. http://www.kba.de/DE/Statistik/Fahrzeuge/fz_methodische_erlaueterungen_201401_pdf.pdf?__blob=publicationFile&v=11. Accessed 06 July 2016
Federal Motor Transport Authority (2015b) Neuzulassungen von Kraftfahrzeugen und Kraftfahrzeuganhängern—Monatsergebnisse September 2015. http://www.kba.de/SharedDocs/Publikationen/DE/Statistik/Fahrzeuge/FZ/2015_monatlich/FZ8/fz8_201509_pdf.pdf?__blob=publicationFile&v=2. Accessed 06 July 2016
Federal Motor Transport Authority (2015c) Pressemitteilung Nr. 01/2015—Fahrzeugzulassungen im Dezember 2014—Jahresbilanz. http://www.kba.de/DE/Presse/Pressemitteilungen/2011_2015/2015/Fahrzeugzulassungen/pm01_2015_n_12_14_pm_komplett.html?nn=1206172. Accessed 06 July 2016
GoingElectric—Elektroauto News (2016) Overview of the current offer of e-vehicles in Germany. http://www.goingelectric.de/elektroautos/. Accessed 06 July 2016
Golloch R (2005) Downsizing bei Verbrennungsmotoren. Springer, Berlin

Ifeu—Institut für Energie und Umweltforschung Heidelberg GmbH (2009) Kurzgutachten Wasserstoff- und Stromspeicher in einem Energiesystem mit hohen Anteilen erneuerbarer Energien: Analyse der kurz- und mittelfristigen Perspektive. http://www.bmub.bund.de/fileadmin/bmu-import/files/pdfs/allgemein/application/pdf/ifeu_kurzstudie_elektromobilitaet_wasserstoff.pdf. Accessed 06 July 2016

Ökoinstitut e.V. (2011) Autos unter Strom. http://www.oeko.de/fileadmin/pdfs/oekodoc/1283/2011-413-de.pdf. Accessed 06 July 2016

Press and Information Office of the Federal Government (2012) Transcript of the press conference after the summit meeting for electromobility of October 1, 2012. http://www.bundesregierung.de/ContentArchiv/DE/Archiv17/Mitschrift/Pressekonferenzen/2012/10/2012-10-01-pk-elektromobilitaet.html?nn=437032. Accessed 06 July 2016

Tesla Motors, Inc. (2013) Blog post „Model S fire". http://www.teslamotors.com/blog/model-s-fire . Accessed 06 July 2016

Tesla Motors, Inc. (2016) Model S. http://www.teslamotors.com/de_DE/models. Accessed 06 July 2016

Toyota Deutschland GmbH (2016a) Toyota Prius Hybrid. http://www.toyota.de/automobile/prius/in-dex.json. Accessed 06 July 2016

Toyota Deutschland GmbH (2016b) Toyota Prius Plugin. http://www.toyota.de/automobile/prius-plugin/index.json. Accessed 06 July 2016

Volkswagen AG (2016a) Der Golf — Technik und Preise. http://www.volkswagen.de/content/medialib/vwd4/de/dialog/pdf/golf-a7/preisliste/_jcr_content/renditions/rendition.download_attachment.file/golf_preisliste.pdf. Accessed 06 July 2016

Volkswagen AG (2016b) Preisliste e-Golf. http://www.volkswagen.de/content/medialib/vwd4/de/dialog/pdf/golf-a7/egolf_preisliste/_jcr_content/renditions/rendition.download_attachment.file/e-golf_preisliste.pdf. Accessed 06 July 2016

Chapter 2
Network Access Pursuant to § 20 EnWG

Against the backdrop of a growing number of e-vehicles and a user-friendly as possible access to public charging infrastructure the following statement shall be verified first:

> ***Electricity suppliers have a right to non-discriminatory access to publicly accessible charging infrastructure in accordance with § 20 EnWG.***

2.1 The Benefits for the End Consumer

There would be multiple advantages for the driver:

- The driver could freely choose a supplier and thus also ultimately the energy product (e.g. a green energy or flatrate product).
- Competition for the supply of the respective vehicles would be enabled and a consumer-friendly development of products, e.g. in regard to the price or other characteristics, could be expected.
- The most important advantage from a customer perspective would be the availability of all public charging stations to charge their vehicles. Based on the respective grid usage contracts, the electricity supplier could supply all charging stations accessible to the public and the vehicle driver would then need only *one* charge card, instead of many, to access and use all nationwide charging points.

Such right to access could already be established under the current Energy Industry Act[1] (EnWG).

[1]Energiewirtschaftsgesetz vom 7. Juli 2005 (BGBl. I S. 1970, 3621), das zuletzt durch Artikel 311 der Verordnung vom 31. August 2015 (BGBl. I S. 1474) geändert worden ist.

M. Adam, *Accelerating E-Mobility in Germany*, SpringerBriefs in Law
DOI 10.1007/978-3-319-44884-8_2

2.2 Network Access Pursuant to § 20 EnWG

In regard to the question, whether a right of network access to public charging stations is established, a brief history of the power supply and the system of network access to the electrical grid shall be presented in order to then investigate as to whether the charging stations are part of the electrical distribution network or whether they are at least an infrastructure similar to the grid.

2.2.1 A Brief History

The issues regarding network access have been discussed in basic form in the past in connection with the liberalization of the electricity market. These experiences could also possibly provide insights for the present question of network access to charging stations.

The starting signal for the regulation of the energy market in Germany should have been the Law on electricity and gas supply (Energy Indusrty Act—EnWG), entered into force on April 24 1998. The Energy Industry Act in force until April 24 1998 was originally entered into force in 1935 and led from today's perspective to a fragmentation of the energy market in Germany—a development that is seen and felt in the form of the plurality of power distribution systems today.

Electrification began in Germany late 19th century, when industrial companies built the first block installations that were able to supply blocks or even smaller streets with electricity.[2] Thus island-like, small-scale networks were formed, which allowed the power supply in the local area around the respective power plant.[3] The electricity sector got a boost by the demand of the municipalities, mainly caused by the electrification of transport in the form of trams and the replacement of the gas lantern by the electric light bulb in streetlights.[4] Municipalities were, as large purchasers of electricity and as owners of the monopoly routes for laying the power lines, the major stakeholders in the German electricity sector in the early 20th century.[5] Thus, already at the beginning of the 20th century a structure of power grids had been created, spatially confined to municipal boundaries and protected by exclusive agreements. A concentration partially took place in rural areas, where cross-country centers developed, which could provide power via three-phase lines not only locally, but also over many kilometers distance.[6] The Energy Industry Act of 1935 protected this established structure of hundreds of power distribution

[2]Evers (1983, p. 23).

[3]Evers (1983, ibidem).

[4]Löwer (1989, p. 35).

[5]Löwer (1989, p. 37); Evers (1983, p. 27).

[6]Mengers (2007, p. 20).

networks. The preamble to the Energy Industry Act 1935 mentioned in addition to the inexpensive and safe energy supply as other objectives of the Act:

> *"... To secure the necessary public influence in all matters of energy supply, prevent harmful effects of competition to the national economy..."*

The power supply was then still regarded as a uniform process that is to say, there was no separation between generation, distribution and sales, so that the fear that competition in this area could be harmful (in the form of redundant infrastructures), in principle is comprehensible. Competition was legally and factually made impossible by so-called demarcation agreements. These contracts secured a particular area to a single energy provider and protected the energy provider from competing companies.[7] The German Competition Law explicitly stated an exception of these contracts by the antitrust provisions in § 103 GWB (version of the year 1958). The result of this more than a hundred year old practice was a variety of local distribution systems, which constitute natural monopolies. This structure is still visible, see Fig. 2.1.

In 2000, more than 900 distribution network operators were active in Germany,[8] and in 2014 there were still 884 distribution network operators who were registered by the Federal Network Authority.[9]

The *Directive 96/92/EC of the European Parliament and of the Council of 19 December 1996 concerning common rules for the internal market in electricity* laid the foundation for the liberalization of the electricity market in Germany. The provisions of the Directive were implemented in the form of the Law on electricity and gas supply (Energy Industry Act—EnWG) by Germany on 24 April 1998. The Energy Industry Act of 1998 meant a paradigm shift, not least because in § 6 EnWG the right of non-discriminatory network access was established for the first time. This marked the end of the unity of electricity grid and electricity sales, virtually a century-old practice.

The new legal framework led to a variety of start-ups in the field of energy, all electricity suppliers without their own network.[10] However, Germany refrained initially from a sector-specific regulation of network access and opted for the variant of negotiated network access, see. § 6 EnWG of April 24, 1998. It was the approach via the so-called association agreements[11] to develop an industry-recognized solution. The association agreements contained in particular provisions for the calculation of network charges. It is noteworthy that the association agreements were

[7] Senke (2003, p. 54).

[8] European Commission (2001, p. 42).

[9] Federal Network Agency and Federal Cartel Office (2014, p. 25).

[10] i.a. LichtBlick GmbH, Riva Energie AG, ares GmbH and Yello Strom GmbH.

[11] A compact description can be found in Seeliger (2000, p. 661).

Fig. 2.1 Map of the distribution networks, lower voltage level—Germany 2016 (*Source* Enet GmbH, https://www.enet.eu/kartestromnetzbetreiber.html; retrieved June 15, 2016.)

essentially developed and adopted without the involvement of network users, i.e. the people affected by the regulations.[12]

The results were devastating: The overpriced network charges caused a prohibitive effect in the direction of new suppliers and in general, the new players in the market for power supply were facing numerous market access obstacles.[13]

The Federal Cartel Office and the civil courts were the only points of contact for the new entrants and the incumbent power companies, who were the owners of the respective electricity grids, were very creative when it came to grounds for refusal or market obstacles. Thus, for example, it was required that prior to changing the energy supplier, costly measuring devices to be installed[14] or a change fee was charged for each change of supplier[15] and in general customers were bothered by the network operator with unnecessary additional contracts.[16] Also, a refusal of network access due to the "lignite protection clause", which should protect the production and sale of electricity from lignite, was a variant of the barrier to market access.[17] Although the Federal Cartel Office committed itself early on and took strong action against these abusive practices, it showed that the ex-post control of Competition Law was ill-suited to provide non-discriminatory access conditions to the new entrants. According to the EU benchmarking report the Federal Cartel Office received around 150 complaints in the first two years of formal market liberalization (and the complaints at the state level are not included in this figure).[18]

In addition to the already cited abusive practices, there was a further variant, which was a major hurdle for new suppliers: The energy supply of final consumers by way of *"resale"* through the incumbent energy provider *("Beistellung").*[19] The Federal Network Agency defined *"resale"* as follows:

> *"In case of resale the incumbent utility supplies a customer on behalf of the new supplier who buys the energy at the incumbent supplier in order to sell it to the final customers. The competitor, the new supplier, stipulates this with the incumbent supplier in form of a resale agreement."*[20]

[12]In addition to the network operators, however, the VIK Association of Industrial Energy and Power Economics e.V. (VIK) was involved in the association agreements. The VIK also represents large network users, mainly connected to higher voltage levels.

[13]Federal Network Agency and Federal Cartel Office (2003, pp. 17–18); BT-Drucksache 15/3917, p. 83; Säcker (2014, pp. 18–19).

[14]Federal Cartel Office (2000a).

[15]Federal Cartel Office (2000b).

[16]Federal Cartel Office (2001).

[17]Regional Court Berlin, decision on June 27, 2000, case no. 16 O 652/99, in: ZNER 2000, pp. 142–150, p. 145.

[18]European Commission (2001, p. 17).

[19]„Resale", a term coined by the Federal Network Agency as a translation for the german word „Beistellung", can be described as the single buyer model laid down in Art. 18 of the repealed Directive 96/92/EC of the European Parliament and of the Council of 19 December 1996 concerning common rules for the internal market in electricity.

[20]Federal Network Agency and Federal Cartel Office (2013, p. 301).

This form of supply to customers excludes procurement and direct supply of electricity by third party suppliers, because the energy supplied is—as before—procured and provided by the incumbent supplier.[21] Rather, the *"resale"* is only a settlement between the incumbent supplier and the new supplier, which only formally gives the customer the opportunity to choose the contractual partner for energy supply. The new supplier is in this "resale system" reduced to a pure bill recipient, because the physical delivery and the actual supply of energy are undertaken via the established and integrated utility, usually the old supplier of the customer. In this "resale" scenario the new supplier is barred from maintaining its own balancing group management and cannot procure energy on the wholesale market. In regard to the upstream markets, here: the wholesale market and the market for the first generation of energy, the newcomers thus remain blocked. Furthermore, there is the latent risk of a price-cost squeeze ("margin squeeze") for the new supplier, since the old supplier can (over)price the "resale" in such a manner that there remains no profit margin for the new supplier.

The "resale system" (or single buyer model) thus made entry into the market for the supply of retail customers more difficult while simultaneously preventing access to the upstream markets.

In particular at the European level all these developments were seen critically and as a result the system of negotiated network access has been abolished in the second internal market package[22] and the creation of a regulatory authority was made mandatory.[23] The Member States were obligated to implement the Directive into national law on 01 July 2004. On 15 July 2005 the German Energy Industry Act and associated ordinances came into force, in particular regulating the right to non-discriminatory network access in § 20 EnWG and the related ordinance (StromNZV). Thus, the attempt at self-regulation, in terms of the negotiated network access, had failed.

2.2.2 The Basic Structure of Energy Supply in the EnWG

With the Directive 2009/72/EC, an essential part of the Third Internal Market Package, the recent regulatory efforts to create a single internal market for energy were further refined and improved. Priorities of the new regulations were particularly

- better unbundling of generation, distribution and sales,
- the strengthening of consumer rights,
- and the duty to introduce smart meters by 2020.

---

[21]Bartsch (2002, p. 105).

[22]Art. 20 and 21 of Directive 2003/54/EC.

[23]Art. 23 of Directive 2003/54/EC.

In the area of network access regulation, the structure remained essentially the same, but for the first time the concept of the customer facility[24] has been legally defined and thus offers a legal differentiation between the publicly accessible network and the private customer facility in the EnWG of 2011.[25]

The provisions for the unbundling of generation, supply, and distribution, as well as the right to non-discriminatory network access, have created a system with clearly separated market roles. The vertical integration of the various value-added stages is thus canceled in principle, although it must not go unnoticed that the vast majority of German utilities (812 of 884) are excluded from the unbundling provisions due to the de minimis provision in § 7 para. 2 EnWG.[26]

The supply networks have now been subject to nearly 10 years of regulation and although the debate about the effectiveness of the regulations has not yet fallen silent, it can be said, but as an interim conclusion, that market power has declined at the production and wholesale level[27] and that at the retail level of energy supply with an average of 97 suppliers per network area a large number of competitors[28] can be found, even if the current change rate of household customers is at only 5.5 %,[29] and as such is certainly in need of improvement.

In other words, the regulation of the network and the implemented system of third party network access are working well.

2.3 The Public Charging Station in the System of the EnWG

The right of electricity suppliers to non-discriminatory network access to charging stations pursuant to § 20 para. 1 EnWG requires various constituent elements whose existence is to be examined hereinafter.

2.3.1 Publicly Accessible Charging Station

First of the all, the object of study is to be narrowed down and defined, because the charger accommodated in the garage at home will clearly be subject to other provisions as for example the charging station on a city-owned parking lot. In Art. 2 no. 7 of Directive 2014/94/EU a legal definition is given:

---

[24]See § 3 no. 24a EnWG.

[25]BT-Drucksache (2011)

[26]Federal Network Agency and Federal Cartel Office (2014, p. 25).

[27]Federal Cartel Office (2011, p. 20).

[28]Federal Network Agency and Federal Cartel Office (2014, p. 19).

[29]Federal Network Agency and Federal Cartel Office (2014, p. 148).

"'recharging or refuelling point accessible to the public' means a recharging or refuelling point to supply an alternative fuel which provides Union-wide non-discriminatory access to users. Non-discriminatory access may include different terms of authentication, use and payment;"

In the explanatory memorandum of the draft for the German charging stations regulation,[30] which is intended to implement the provisions of Directive 2014/94/EU into national law, a paraphrase of the characteristic "public" is found, which has already been proved useful in § 35 VwVfG to define and make a distinction between the terms "public" and "non-public". A charging station is deemed accessible when it is addressing the general public and just not *"directed at special a group of persons, which is identified or identifiable by general characteristics."*

The definition thus takes into account the will of the operator: If the charging station is directed at an indefinite multitude of individuals, it is publicly accessible. However, if the charging station is only aimed at individual persons or a determinable group of people, public accessibility cannot be assumed. Thus, the tedious distinction between public and semi-public charging facilities is particularly obsolete in view of the possibility of access. Rightly, a charging device is publicly available or not—regardless of whether it is located on private or public ground.

2.3.2 Rightful Claimant

In accordance with § 20 para. 1 sentence 1 EnWG anyone, i.e. any natural or legal person (network users in accordance with § 3 no. 28 EnWG), is entitled to grid access. In § 20 para. 1a EnWG it is clarified that this right is granted both to the end consumer who wants to be supplied, as well as the energy supplier who wants to undertake the supply. As long as an energy supplier submits a claim to access a charging station the supplier will therefore be entitled to a right of network access on its merits.

2.3.3 Obligated Party

The claim to network access is brought against the operator of the electricity supply system, raising the central question of whether charging station operators are deemed to operate an electricity supply system, and thus charging stations are to be classified as electricity supply system or possibly as a part of such electricity supply system.

The term "operator of an energy supply network" encompasses pursuant to § 3 no. 4 in conjunction with § 3 no. 2 EnWG especially operators of electricity

[30]Federal Ministry for Economic Affairs and Energy (2015b).

distribution networks. Distribution system operators are defined in § 3 Nr. 3 EnWG as a company, "*that carry out the function of distributing electricity and are responsible for the operation, maintenance, and, if necessary, development of the distribution system within a given area and, where applicable, interconnection to other networks.*"

This definition is supported by the definition in § 3 no. 16 EnWG, according to which an energy supply system is defined as a electricity supply system (or gas supply system) through one or more voltage (or pressure) levels, however, with the exception of customer facilities pursuant to § 3 no. 24a and 24b EnWG, see § 3 no. 16 EnWG.

The EnWG surprises to the extent that, although there are numerous definitions of terms in § 3 of the Energy Act, no definition of this very central and important term of electricity supply system is given. The same applies to Directive 2009/72/EC, which also does not provide any definition.

Therefore, the classification of the charging infrastructure in the system of the EnWG will depend on an overall view and in particular suggests a study and delineation of the various facility groups in the EnWG.

Undoubtedly, the charging station can be classified as an energy system within the meaning of § 3 no. 15 EnWG,[31] because a charging station is used to supply energy to an electric vehicle. While this may initially be considered a rather banal observation, this association does develop a rather far-reaching effect, since it clarifies that charging stations fall within the scope of the EnWG. In particular, the construction and operation of charging stations must therefore be measured against the goals of § 1 EnWG.

In addition to the classification as an electricity supply system, the subsumption as a direct line, closed distribution network, customer facility or facility sui generis, i.e. as a facility that is not defined in the Energy Industry Act, should be examined as well hereinafter.

2.3.3.1 Direct Line

The technical structure of a charging station suggests a classification as a direct line, because the charging station actually consists of a charging column and a (power supply) line which connects the public electricity network with the charging station. The definition in § 3 no. 12 EnWG describes a direct line in the second alternative as a "*line which connects an electricity producer and an electricity supply undertaking to their own business unit, subsidiaries or customers for the purpose of direct supply.*" Unlike the directive 2009/72/EC, which ensures the customer's right to free choice of supplier even in the case of using a direct line,[32] no third party access would be possible with a qualification as direct line according to the

[31]The Draft for the German charging stations regulation uses the same approach.

[32]See Art. 2 no. 15 in conjunction with Art. 33 and Art. 34 of Directive 2009/72/EC.

current EnWG because under German law the direct line is excluded from the right to network access pursuant to § 20 EnWG. However, it is highly doubtful that this is permissible under European law, especially in light of the „*effet utile*".[33] But ultimately it also lacks the requirements of § 3 no. 12 EnWG, since there will regularly not be a connection between producers and electricity supply undertakings in the case of a charging station, but rather a connection to the general supply system. In the case of charging stations, the purchased electricity is not supplied directly by a producer, but by a supplier via the balancing group system, see § 20 para. 1 EnWG in conjunction with § 4 StromNZV.

2.3.3.2 Closed Distribution System

The closed distribution system in accordance with § 110 EnWG may seem at first glance as a possible alternative for the classification of charging stations, as closed distribution systems have similarities with publicly accessible charging stations because of their small size in technical terms. However, the conditions for the existence of a closed distribution system will almost never be met. It follows in particular from § 110 para. 2 EnWG that over the closed distribution system no or only a small number of final costumers will be supplied. However, publicly accessible charging stations are aimed precisely at the largest possible number of final customers.

Incidentally, even if charging stations were classified as closed distribution system, a right of network access would be established since § 20 EnWG is not part of the exempted provisions for closed distribution systems, listed in § 110 para 1 EnWG.

2.3.3.3 Customer Facility

With the aforementioned amendment to the Energy Industry Act in 2011, the concept of the customer facility was legally defined the first time. The classification of public charging stations as customer facilities within the meaning of the Energy Industry Act was hitherto probably the prevailing opinion in the literature.[34] But when the legislator in this amendment to the Energy Industry Act made clear that third party access has to be granted in customer facilities (which is even provided free of charge in contrast to the right of network access pursuant to § 20 EnWG), the supporters of this interpretation became silent.[35]

[33] For the legal concept of „effet utile" and further sources see Potacs (2009, pp. 465–487).

[34] Fest et al. (2010, p. 93); Feller et al. (2010, p. 240).

[35] Left open by Lüdemann et al. (2014, p. 6); Apparently given up by de Wyl (2015); Also given up by Claus Fest, who voiced the opinion, that public charging stations are facilities sui generis, to the author in an expert talk in Berlin on April 29, 2014.

The conceptual description in § 3 no. 24a EnWG, according to which customer facilities are defined as energy systems to supply energy, initially suggests a classification of charging infrastructure as a customer facility. It is undoubtedly an energy system, and a supply of energy—in the form of electricity, which is delivered to electric vehicle drivers—also takes place. However, the other conditions specified in subparagraphs (c) and (d) of the legal definition ultimately rule out the classification of charging stations as customer facilities. On the one hand the charging station must then be insignificant for ensuring effective and undistorted competition in the supply of electricity, see § 3 no. 24 (c) EnWG. Looking at the still low penetration of charging stations and electric vehicles one may be inclined to classify the charging stations as insignificant. However, even small e-vehicles with an annual mileage of 10,000 km/a consume as much as a small household, namely 1,500 kWh per year and more. With an intended target of one million vehicles in 2020[36] and 150,000 public charging stations by 2020,[37] it is thus of rather high importance for ensuring effective and undistorted competition in the supply of electricity to be asserted. For the energy sector, it means accessing a completely new target group—the e-vehicle drivers—for the supply of electricity, and to attract them as new customers. The criterion of the free network access in subparagraph (d) of the definition is in the end contrary to a classification of the charging station as a customer facility, as payment is regularly demanded for using the charging station, but a network access for third party suppliers is currently not granted.[38] The explanatory memorandum of the EnWG, BT Drucksache 17/6072, p. 95, reiterates the right of third party network access in regard to customer installations, since resale (or single buyer) models are explicitly prohibited.

The definition in § 3 no. 24 EnWG and the associated legislation materials suggest therefore that the publicly accessible charging station is not a customer facility. But even if the publicly accessible charging station were to be classified as a customer facility that would mean, however, that a third party access must be granted.

2.3.3.4 Facility Sui Generis

The history of the EnWG commends for the classification of publicly accessible charging stations as facilities sui generis, i.e. a facility that is not defined in the Energy Industry Act,[39] since publicly accessible charging stations have not been defined in detail, as these facilities have not yet played a significant role in the public debate, especially when the larger amendments to the Energy Industry Act in 2006 and 2011 were discussed and adopted. But with the latest amendment in 2014,

---

[36] Federal Ministry for Education and Research (2009, p. 2).

[37] European Commission (2014).

[38] The only exception here was the so-called „Hamburger Model“, see Sect. 2.3.5.

[39] Heinlein (2013, p. 219).

the publicly accessible charging station is now mentioned in § 49 EnWG and § 53b no. 1 EnWG. Above all, the explanatory memorandum to § 53b no. 1 EnWG leads to the assumption that the legislature actually assigned the charging station to a non-specified category of energy systems.

The sentence

> *"To this end, it is useful to include the charging infrastructure at a main spot in the general system registry and provide this information to the system operators."*

distinguishes the charging station from the energy supply system, so that on this reading the publicly accessible charging stations are indeed energy systems, but not customer facilities and also not part of the power grid. At a first glance, the subsumtion of a charging station as a sui generis installation therefore stands to reason.

But does the lack of an explicit definition warrant an exception to the energy regulation? In light of the legal requirements for customer facilities, that need in particular to grant third party network access free of charge, it appears, a fortiori, at least doubtful, to want to exclude publicly accessible charging stations from energy regulation.

2.3.3.5 Energy Supply System

A frequently used definition describes the energy supply system as a system of power lines, which are interconnected and intermeshed, and do not solely serve the aim to supply individual customers with energy.[40] The Federal Court of Justice has rejected this definition as too restrictive, since even single feeder lines, for example, fall within the scope of the definition of an energy supply system.[41] That even the smallest grids are subject to the concept of an energy supply system and a various number of connection points is explicitly not needed for the classification of an installation as an energy supply system, is the standing jurisprudence.[42] As a result, even the Federal Court of Justice avoids in the cited decision of 18 October 2011 a final definition and provides rather that the concept of the energy supply system should be seen from the customer's perspective, i.e. the person who is supplied with electricity.[43]

Both the Directive 2009/72/EC and the EnWG provide a definition of the final customer, see Art. 1 pt. 9 Directive 2009/72/EC, or the final consumer, see § 3 no. 25 EnWG, which can be used as a starting point to narrow down, in a regulatory sense, the charging of an electric vehicle at a public charging station. According to

[40]Salje (2006, p. 245), recital 201.

[41]FCJ, Decision on October 18, 2011, case no. EnVR 68/10, recital 19.

[42]Higher Regional Court Düsseldorf, Decision on April 05, 2006, case no. VI-3 Kart 143/06 (V), recital 24f.; FCJ, Judgement on November 10, 2004, case no. VIII ZR 391/03, recital 15; ECJ, Judgement on May 22, 2008, C 439/06 („Citiworks"), recital 49.

[43]FCJ, Decision on October 18, 2011, case no. EnVR 68/10, recital 21.

the definitions referred to, final customers or final consumers are people or customers purchasing energy for their own consumption. This applies exactly to the driver that charges his vehicle at a public charging station: The electrical energy that is supplied by the charging station is used within the vehicle to be converted into kinetic energy and other forms of energy (light, heat, etc.). There is therefore a final consumption by the driver of the electric vehicle, because the electricity is not transmitted elsewhere, but used in the vehicle—the electricity is consumed. The operator of the charging station is therefore no final consumer, because there is no consumption by the charging station itself (except for powering the station), but a distribution and supply of electricity (meaning the sale of electricity).[44]

In regard to the definition of the final consumer in Article 2 Nr. 9 of Directive 2009/72/EC, which provides that "final customer" means

„a customer purchasing electricity for his own use,"

the incompatibility of the envisaged amendment to § 3 no. 25 EnWG in the draft bill of the "Electricity Market Law"[45] with Directive 2009/72/EC is proven incidentally. According to this draft bill, electricity purchases for charging points are to be considered as final consumption in the future.[46] However, this is, as described, readily recognizable as in contradiction with the actual practice whereby charging stations and charging points do precisely not consume electricity themselves, but rather this electricity is distributed and supplied to final consumers—the drivers of electric vehicles.

With finding that charging stations ultimately supply final customers, the classification of the charging station as a closed distribution system must be ruled out, because the supply of final customers is an exclusion criterion in accordance with § 110 para. 1 EnWG.

At first glance the small physical size speaks against the classification as an energy supply system, or as a part of such system. However, from a systematic point of view it does not seem advisable to focus solely on the size of the technical device. Neither German law nor the European Directive know a minimum size for electricity networks, but the legal requirements focus rather on the importance of the infrastructure for the competition. A classification as energy supply system therefore seems possible in principle. As a result, this raises the question of whether the charging station can be defined as an energy supply system, a customer facility or a facility sui generis.

---

[44]Rejecting: Feller et al. (2010, p. 243).

[45]Federal Ministry for Economic Affairs and Energy (2015a).

[46]Federal Ministry for Economic Affairs and Energy (2015a, b, ibidem, p. 9).

2.3.3.6 Discussion: The Legislative Purpose

A classification of the publicly accessible charging infrastructure must be based on legislative purpose, eventually. Pursuant to § 1 para. 1 EnWG, the legislative purpose "*is the most secure, low-priced, consumer-friendly, efficient, and environmentally compatible grid-bound supply of electricity and gas for the general public, which is based increasingly on renewable energies.*"

Supply of Electricity for the General Public

The publicly accessible charging infrastructure serves directly to supply motorists with electricity. The charging infrastructure is also addressing an indefinite number of motorists. That means, that publicly accessible charging stations offer a grid-bound[47] supply of electricity to the general public. However, whether a public charging station can be classified as a customer facility, an energy supply system, or an facility sui generis, cannot be solely assessed on this formula.

Consumer Protection

The idea of consumer protection in § 1 para. 1 EnWG, which has undergone numerous concretisations especially in §§ 40 to 42 EnWG, takes the discussion further though:

The EnWG determines in § 40 para. 1 EnWG that invoices for energy supplies to final costumers must be simple and understandable. Concrete specifications for the invoice contents are made in paragraph 2, e.g. the determined energy consumption for the billing period must be shown on the invoice and for household customers the initial and final count of the energy metering must be shown in kilowatts per hour in accordance with clause 4.

The invoicing of numerous charging stations is, however, currently done in EUR per minute of charging, i.e. the costumer is not billed the amount of energy, as would be usual in the case of electricity supply. As a result, the bills are difficult to understand and the prices are, in some cases, very high.[48] There are virtually no plausible reasons given at all, why exactly the charging of electric vehicles should not satisfy the same consumer protection requirements, as any other power consumption. In regard to electricity labelling, which is compulsory in § 42 EnWG, the same thinking should be applied: The charging stations operator will probably be able to show the electricity labeling directly to the customer, provided that a contract has been explicitly concluded. However, if the costumer receives his

[47]Cf. Federal Network Agency in Decision of May 20, 2014; case no. BK7-13-073, p. 12.

[48]See for example the report by ZDF-programme „WISO" (2015).

invoice via SMS or via a third party invoice, based on a roaming agreement, that information will usually not be shown.

But the electricity mix is a key product feature that is of high importance particularly in the area of e-mobility and a waiver of this transparency provision is therefore unacceptable. This is all the more the case since the legislative objective formulates an increasing supply through renewable energies as a main target.

Therefore this legislative objective would be thwarted by a lack of electricity labelling.

Effective and Genuine Competition

Another objective is stated in § 1 para. 2 EnWG, according to which the regulation of electricity and gas supply networks

> "*serves the goals of ensuring effective and genuine competition in the supply of electricity and gas and ensuring efficient and reliable operation of energy supply systems for the long-term.*"

The publicly accessible charging stations exclusively use the resale model (or single buyer model) and a supply of electricity from third-party suppliers via a third party network access is currently not practiced.[49] As already described in Sect. 2.2.1, the resale model rather prevents competition, as foreclosure of the upstream market (here: the wholesale electricity market) is the result and the resale model is prone to abusive behaviours in the form of „*margin squeezes*".

Thus it is shown, that there is currently no effective and undistorted competition, but rather the risk of restricted competition. Taking the idea of competition into account thus inevitably leads to a classification of charging stations as part of the regulated energy supply system.[50]

However, this would raise the legitimate question whether this constitutes a breach in the current statutory system, since the legislature made a conceptual separation of energy supply system and charging infrastructure in § 53b EnWG.

That interpretation can be countered in that way, however, that a conceptual distinction in regard to the general system registry regulated in § 53b EnWG, does not warrant a different treatment regarding the main regulatory categories such as network access and tarification.

---

[49] A third party network access is no longer being granted at public charging stations in the Hamburg distribution network since 2015. Staff members of Stromnetz Hamburg GmbH, the operator of the public charging infrastructure in Hamburg, confirmed this to the author during an expert talk in Hamburg on January 26, 2015.

[50] Consenting: von Hoff (2009, pp. 342–343), Weis (2014, p. 288), Hartwig (2013, p. 482), Schau and Götz-Friedrich (2007, p. 124), Reinke (2014, p. 175); rejecting: Lüdemann et al. (2014, p. 6), Böwing (2013, p. 97), Boesche (2013, pp. 188, 201) , Keil and Schmelzer (2010, p. 566), Michaels et al. (2011, p. 831).

Rather, the legislature stresses in the explanatory memorandum to § 14a EnWG, which also relates to charging devices of electric vehicles, that the charging and discharging of electric vehicles is essentially a competitive process.[51] Therefore the EnWG amendment from August 04 2011 was a clear manifestation of the legislature for a competitive classification of charging devices. If the non-public charging points are part of the competition, then what could possibly be the argument that publicly accessible charging points should not be part of the competitive supply? This *"a fortiori"* argument thus leads to a classification of the publicly accessible charging points as part of the competitive energy market and thus as part of the energy supply system.

Historically, the Energy Industry Act was created for that very reason, to provide the final consumer effective choices when choosing their electricity supplier. However, the current situation in regard to charging stations is strikingly similar to the state of electricity supply since liberalization in 1998 and the first few years thereafter, where the free choice of supplier was formally enabled, but the resale model and other market barriers almost brought competition to a halt.

The current market development of charging stations—roaming agreements as an attempt at self-regulation, the resale model as a feigned competitive solution—therefore presents itself as an almost exact repetition of the period from 1998 to 2005 as described in Sect. 2.2.1. It is highly doubtful that this development corresponds to the will of the German and especially the European legislator.

Thus it has already become apparent that the classification as a facility sui generis, which would be subject to the EnWG in general, but not subject to the specific regulation of competition by the EnWG, has no basis at all.

Since a customer facility by legal definition in § 3 no. 24a EnWG only exists if it is insignificant for competition, the publicly accessible charging station will certainly not be a customer facility. The charging infrastructure shall precisely serve an unspecified large number of customers—according to the will of the commercial operator to the largest possible number of customers—so that it certainly is a competition-related infrastructure. This becomes even more apparent in regard to the objective of the Federal Government of 1 million electric vehicles by 2020:[52] With an annual mileage of 10,000 km per year and an average consumption of around 15 kWh per 100 km[53] an e-vehicle therefore consumes approximately 1,500 kWh per annum. Over 35 million Germans lived in a rented accomodation in 2013[54] and will therefore not be equipped with charging installations at home in the future. A significant proportion of charging will therefore necessarily take place at publicly accessible charging stations. The fact that this is a competition-related

[51]BT-Drs. 17/6072, p. 138.

[52]Confirmed by the Federal Ministry for Eocnomic Affairs and Energy on April 07, 2015, see Zeit Online (2015).

[53]For the BMW i3 a gross consumption of 16,7 kWh/100 km is stated, ADAC (2013, p. 12), https://www.adac.de/_ext/itr/tests/Autotest/AT5053_BMW_i3/BMW_i3.pdf.

[54]Statista GmbH (2015).

field, can be shown by comparing the issue of charging stations to the comments of the Federal Court of Justice in the decision of 18 November 2011, case no. EnVR 68/10. In this case, a campsite, hosting tenants with an annual consumption ranging from 50 kWh to 2300 kWh was object of the proceedings.[55] This is ultimately the order of energy consumption, which is supplied per annum by charging stations. Fortunately, the Federal Court of Justice answered in the cited decision the question of the temporary-only connection of final costumers, which happens also to arise at charging stations. So it says in paragraph 20:

"The finding, that the concerned installation is therefore an energy supply system and not an accumulation of direct lines, is not altered by the fact that some tenants connect to the network by plug-in connections. Because it does not matter how the connection to the grid is made, but that the tenant is connected at all to the energy supply system of the herein concerned party."

This assessment applies similarly to charging stations.

According to the jurisprudence of the FCJ, however, it should be less the technical characteristics, but rather the protection of competition, which is decisive for the classification as an energy supply system.[56] The Higher Regional Court of Stuttgart very clearly formulated this in the previous instance:

"Because the complainant monopolizes the energy supply and excludes, in contradiction to the objectives of the Energy Industry Act, the possibility of the users of choosing their suppliers and therefore of choosing the conditions of their own electricity purchases themselves and thus ensure competition in energy supply."[57]

The described situation is absolutely comparable to the publicly accessible charging points: There is no possibility for the consumer to choose a supplier.

The FCJ has confirmed its position on the broad interpretation of the concept of the energy supply system in the decision *"Stromnetz Homberg"* and states very accurately:

„It is in accordance with the main targets of the fundamentally revised Energy Industry Act in 2005, to enable the consumer to choose his electricity supplier. However, this requires a broad understanding of the concept of energy supply. ***In order to enable the supply of electricity by every supplier in principle, all facilities that serve to supply the final costumer, must fall under the term of energy supply system.***"[58]

(Annotation: Highlighting by the author*)*

[55]FCJ, ibidem, recital 17; verbatim however described with „50–2300 KW" (thus stating power data instead of consumption data), but this seems to be a scrivener's error, because the judgement uses consumption data in other instances and of course camping sites normally don't use electric power in the tune of several thousand kW, i.e. the power requirement of a large industrial customer.

[56]FCJ, ibidem, recitals 12 and 16f.

[57]Higher Regional Court Stuttgart, Decision on May 27, 2010, case no. 202 EnWG 1/10, recital 48.

[58]FCJ, Decision on June 03, 2014, case no. EnVR 10/13, recital 35.

The legislative purpose therefore calls for a classification of public charging stations as part of the electricity supply system. Any disturbing feelings, because all the "new" suppliers of public charging facilities have to face the unfamiliar and potentially costly regulation by the Federal Network Agency, are unfounded, since the vast majority of providers are subsidiaries of network operators or in any case subsidiaries of power supply companies.[59] Any difficulties stemming from regulation itself are in principle not unreasonable because they are imposed on all network operators on the one hand by the European single market policy equally and on the other hand any costs can ultimately be allocated to network users.[60]

2.3.3.7 Discussion: Interpretation in Conformity with European Law

The classification of the publicly accessible charging infrastructure must ultimately also be undertaken on the basis of European law, i.e. the interpretation of German law must be compatible with European law.

Applicability of Directive 2009/72/EC

As explicitly mentioned in § 1 para. 3 EnWG, the Energy Industry Act is the implementation and enforcement of European Community law in the field of grid-bound energy supply.

There are hardly any arguments why a publicly accessible charging station should not be part of the grid-based energy supply: Especially in technical terms there has to be a grid-bound supply of electric vehicles with electricity assumed in the case of charging stations. From a structural point of view there is hardly any discernible difference to offshore converter stations. While the converter stations connect multiple wind farms to the transmission network,[61] a charging station connects one or more electric vehicles to the grid. The charging device is so to speak, the access point to the distribution system.[62]

As a result, the operation of publicly accessible charging stations must therefore be governed by the provisions of Directive 2009/72/EC. That interpretation is ultimately supported by recitals 30 and 31 of Directive 2014/94/EU on the deployment of alternative fuels infrastructure:

> *„(30) In the development of infrastructure for electric vehicles, the interaction of that infrastructure with the electricity system, as well as the electricity policy of the Union,* ***should be consistent with the principles established under Directive 2009/72/EC.*** *The establishment and operation of recharging points for electric vehicles should be developed*

[59]See the exemplary list in Annex A—Overview: Operators of Charging Stations.

[60]Federal Network Agency (2010, p. 11).

[61]Tennet Holding (2014b).

[62]See also Peters (2012, p. 153).

as a competitive market with open access to all parties interested in rolling-out or operating recharging infrastructures.(31) ***The access of Union electricity suppliers to recharging points should be without prejudice to the derogations under Article 44 of Directive 2009/72/EC.*** "

(Annotation: Highlighting by the author)

The principles of Directive 2009/72/EC include in particular the unbundling of distribution and power, non-discriminatory network access and an effective level of regulatory supervision, see recital 4 of Directive 2009/72/EC. For the implementation of these principles, it has taken three attempts so far at EU level (Directive 96/92/EC, Directive 2003/54/EC and Directive 2009/72/EC), so the importance of these principles cannot be overestimated.

In addition to the quoted recital 30 of Directive 2014/94/EU, which refers to the principles of the Directive on the internal electricity market, recital 31 of Directive 2014/94/EU is equally revealing, as it refers to the exemptions of the Directive 2009/72/EC in Art. 44. In addition to the country exemptions for Cyprus, Luxembourg and Malta, Art. 44 para. 1 enables the Member States the possibility to apply to the Commission for exceptions for micro isolated systems, e.g. exemption from network access in accordance with Chap. 6. Especially with this reference it is clear that the rules of network access should apply in principle to the publicly accessible charging infrastructure, otherwise the recital 31 of Directive 2014/94/EU would be without any meaning at all.

Applicability of Directive 2009/28/EC

As already shown in Sect. 1.2.2.2, priority access or guaranteed access for renewables to the energy supply system is regulated in Art. 16 of Directive 2009/28/EC on the promotion of the use of energy from renewable sources. In combination with the objective of this Directive, to increase the share of renewables in the transport sector to 10 %, and not to rely solely on biofuels,[63] it must therefore be concluded that network access must also refer to the charging infrastructure; otherwise, this aim of the directive would be jeopardized.

Applicability of Directive 2012/27/EU

A non-discriminatory network access to publicly accessible charging infrastructure can also be derived from the provisions of Directive 2012/27/EU on energy efficiency.

As explained in the section under paragraph 1.2.2.3, the field of e-mobility is covered by the Directive, in so far as the topics "load control" and "smart grids" are named as important components for energy efficiency in recitals Nr. 44 and 45. The

[63]See recitals no. 8 and 18 of Directive 2009/28/EC.

batteries of electric vehicles can be integrated in the electricity supply system via an intelligent control system and thus allow for more efficient management of the electricity supply system. E-vehicles are therefore connected bidirectionally to the grid and may, depending on the load situation in the network, load energy from the grid or feed energy back to the grid (so-called "vehicle-to-grid technology").

The specific rules are laid down in Article 15 of the Directive, which states, inter alia, that there must be non-discriminatory market access by third parties to the markets of ancillary services. Providers of demand response solutions, especially aggregators, are granted such a claim against the operators of the transmission and distribution grids. These future applications in the field of e-mobility can be managed only by means of a non-discriminatory network access. With regard to the planned number of 150,000 publicly accessible charging stations in Germany, this charging infrastructure must not be excluded from providing ancillary services for the distribution and transmission network. Current projects, such as the INEES project sponsored by the Federal Ministry of Economics,[64] show that these vehicle-to-grid technologies are tangible.

Non-discriminatory network access is indispensable for the implementation of such technologies and therefore already established in Art 15 para. 8 of the Energy Efficiency Directive.

Applicability of Directive 2014/94/EU

As already mentioned above, the Directive of the European Parliament and of the Council on the deployment of alternative fuels infrastructure, Directive 2014/94/EU (AFI Directive), provides in recitals 30 and 31 solid evidence that the publicly accessibke charging station is classified as part of the electricity supply system. The reference in Art. 4 para. 7 to smart metering systems as defined in Article 2 point 28 of Directive 2012/27/EU (Energy Efficiency Directive) is another indication that charging stations are to be treated as part of the electricity grid.

However, one must admit that the regulations in Art. 4 para. 8 to 11 of the AFI-Directive at first glance allow for a different way of looking at the classification of charging stations. In Art. 8 it stipulates that charging station operators generally may purchase energy from any energy provider and that the operator of the charging points may also provide services for recharging of electric vehicles to customers. This provision is certainly not an argument against network access to charging stations, but it may be interpreted as derogation from the unbundling provisions of Directive 2009/72/EC, i.e. the separation of network and distribution.

In Art. 9 of the Directive it is stated that also ad hoc charging at any public charging station—i.e. a charging without entering into a contract—must be made possible. Again, this is not an argument against the classification of the charging station as an energy supply system, but it offers customers a new, alternative supply

[64] VDI/VDE Innovation + Technik GmbH (2012), LichtBlick (2012).

option at the charging station—charging without continuing obligation. This scheme is supported by Art. 10, according to which the rates demanded by the charging station operators must be „*reasonable, easily and clearly comparable, transparent and non-discriminatory*." This provision also ultimately represents no exclusion for network access to charging stations. The same applies to Art. 11, according to which distribution system operators have to work together with charging station operators without discrimination.

To this extent the wording of the AFI directive provides both arguments for the classification of publicly accessible charging stations as a distribution supply system, as well as arguments for a different classification of the charging station.

In weighing the arguments, and in view of the provisions already mentioned, there are indications that the AFI-Directive does not want to restrict the operation of charging stations to a specific model and instead provides for the coexistence of network access and direct sale by the charging station operators for the start-up phase. The reference to the principles of Directive 2009/72/EC, which after all is the third version of a directive for the completion of the internal market of electricity, as well as the reference to the Renewable Energy Directive and the Energy Efficiency Directive, both of which also regulate network access for third parties, certainly show that the AFI-Directive is meant to blend in with the existing legal framework. While the aforementioned Directives 2009/72/EC, 2009/28/EC and 2012/27/EU represent long-term objectives of the EU, the AFI Policy is a rather new scheme, which is accordingly equipped with several review clauses. While this does not lead to a recession of the objectives of this Directive, it also does not provide for precedence in relation to the other mentioned Directives.

A look at the historical development of the AFI-Directive shows the intent to provide for a coexistence of business models and, most notably, also to establish network access to the charging station:

In the proposed text of AFI-Directive by the Commission paragraph 14 of the recitals governed:

> „*With respect to recharging points for electric vehicles which are not publicly accessible, Member States should ensure consistency and aim to maximise synergies with intelligent meter roll-out plans following the obligation under Annex I.2 of Directive 2009/72/EC of the European Parliament and of the Council of 13 July 2009 concerning common rules for the internal market in electricity and repealing Directive 2003/54/EC.* ***Publicly accessible recharging points are currently not part of the regulated activities of a distribution system operator as defined in Chapter VI of Directive 2009/72/EC.***"[65]

(Annotation: Highlighting by the author)

This passage was deleted in the final version of the directive and instead of this former provision, the explicit reference to the principles of Directive 2009/72/EC is made in paragraphs 30 and 31 of the recitals, and thus also a reference to the right of third party network access.

[65] European Commission (2013).

The AFI-Directive therefore allows for the classification of charging stations as part of the electricity supply system and in particular provides the right of network access to public charging stations.

2.3.3.8 Interim Result

In regard to the purpose of the Energy Industry Act and the relevant European legal framework publicly accessible charging infrastructure must therefore be classified as part of the electricity supply system.

2.3.4 *Balancing Group*

The electricity supplier must operate a balance group which is part of a balancing group system in accordance with § 20 para. 1a S. 5 EnWG, § 3 para. 2 and § 26 StromNZV. The energy-related processing of trading, distribution and grid activities is performed via a balancing group and balancing groups are virtual energy quantity accounts in order to balance all the actual supplies and withdrawals (physical grid connections) within a control area.[66] Every electricity supplier maintains in practice such a balancing group. A balancing group is managed independently at large market participants and at many smaller energy suppliers the balancing group is managed by a third party provider.

Therefore this requirement will be met regularly.

2.3.5 *No Grounds for Refusal*

Since a right of third party network access to the public charging infrastructure has been proven in principle, the question arises whether there are possible grounds for refusal present in accordance with § 20 para. 2 EnWG. Such a reason may be at hand, if granting access for operational or other reasons is not possible or reasonable, taking into account the objectives of § 1 EnWG.

In the current debate the unreasonableness will be invoked fairly quickly. The lack of appropriate market processes to put network access to charging stations into practice is cited as a key impediment.[67] Thus far there have been two key business processes implemented by the Federal Network Agency with the "regulation of

[66]Tennet Holding (2014a).

[67]Franz and Fest (2013, p. 159).

business processes to supply customers with electricity (GPKE)", case no. BK6-06-009, and the "market rules for the implementation of the invoicing for electricity balancing groups (MaBiS)", case no. BK-06-07-002, which are binding on the whole energy supply sector and govern the interaction between the different parties, such as distribution system operators and electricity suppliers. Contrary to popular belief[68] these business processes can be also applied in their current form to the charging of electric vehicles at public charging stations:

Largely unnoticed by the energy industry and the public, a project funded by the Federal Ministry of Transport took place in the model region of Hamburg, called "hh = more", where non-discriminatory network access to public charging stations has been successfully tested from 2009 on for the first time.[69] Non-discriminatory network access was implemented by Stromnetz Hamburg GmbH, the local distribution system operator, until 2014.[70] As can be seen from the Master Plan for Electric Mobility Hamburg, the implementation ran smoothly[71] and the number of charging processes is impressive (see Fig. 2.2).

Some 7,600 charging processes were handled on the basis of the regulations laid down by the Federal Network Agency standards (GPKE and MaBiS) at public charging stations in Hamburg just for 2013.[72] By the end of the project in 2014 there were more than 13,000 charges that were handled by the standard processes of the Federal Network Agency (GPKE and MaBiS).[73]

Given this project's success, the feasibility and practicality of non-discriminatory network access at public charging stations has been proven.

The developed method was of brilliant simplicity and efficiency:

A virtual metering point[74] was stored on the driver's charging card (a RFID card). When registering at a public charging station, this metering point was transmitted to the charging station operator and used for an assignment of the charging customer (the driver) and his electricity supplier via the back-end of the IT system of charging stations operator. The registration at the charging point was carried out as the process step "delivery beginning", called "moving-in",[75] of the GPKE regulation of the Federal Network Agency, the termination of the charging

---

[68]PricewaterhouseCoopers (2014), Franz and Fest (2013), ibidem; Boesche (2015, p. 449).

[69]Federal Ministry of Transport and Digital Infrastructure (2009).

[70]See Freie und Hansestadt Hamburg (2014, p. 29).

[71]This was confirmed by the responsible staff member of Stromnetz Hamburg during an expert talk with the author in Hamburg on January 26, 2015.

[72]Freie und Hansestadt Hamburg (2014, p. 11).

[73]Ibidem.

[74]A virtual metering point is an alphanumeric combination of characters that is issued only once and allows an unambiguous assignment of the customer within the energy industry business processes, see BDEW (2008, pp. 10–11).

[75]See Federal Network Agency (2011, p. 24).

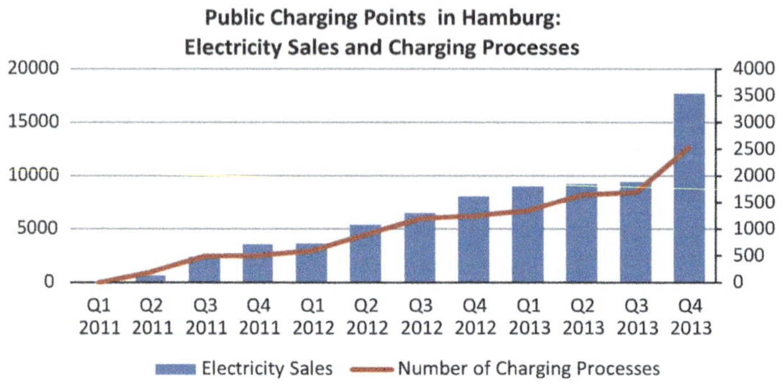

Fig. 2.2 Charging processes at public charging points in Hamburg (*Source* Own diagram, based on Freie und Hansestadt Hamburg, Masterplan zur Weiterentwicklung der öffentlich zugänglichen Ladeinfrastruktur für Elektrofahrzeuge in Hamburg, p.12)

process was accordingly carried out as the process step "delivery end", in the form of a "moving-out". The charging process was implemented at the balancing group level as a so-called flat-rate system.[76] This was recognized with a flat-rate consumption of 1,000 kWh per annum. The classification as a flat-rate system also had the advantage that no special costs for metering were thereby incurred. The data transfer between the energy supplier and charging station operators would depend on the customer's request with regard to the accounting period, see § 40 para. 3 EnWG. When using public charging stations a monthly statement for the sake of traceability of the charging processes is certainly preferable.

This system can be readily duplicated for each network area, i.e. the virtual metering point for each network area would then be stored on the RFID card and the customer could therefore identify himself nationwide with a single card at the respective charging stations and purchase electricity at every charging station from the electricity supplier of his choice.

All in all a robust system that is perfectly suitable for the market launch. In the long run it is certainly appropriate to amend the GPKE processes especially for e-mobility, but as long as the goal of 1 million electric vehicles has not yet been achieved, the energy industry in Germany can easily use this tried and already practiced procedure—the "Hamburg model".

The unacceptability of granting third party network access to public charging stations in accordance with § 20 para. 2 sentence 1 of the EnWG is therefore not feasible.

[76]See § 18 para. 1 s. 2 StromNZV.

2.3.6 *Result*

Electricity suppliers therefore have a right to non-discriminatory access to public charging stations in accordance with § 20 para. 1 EnWG. The claim is brought against the operator of the charging station, who is a system operator within the meaning of the EnWG.

2.3.7 *Excursus: The Tesla Supercharger Network*

The American carmaker Tesla Motors is building its own network of charging stations worldwide, which can only be used by the drivers of the Model S and Model X due to technical characteristics. The DC charging stations have a charging capacity of up to 120 kW.[77] It is therefore by far the fastest and most powerful charging station in the world, which is produced in series. By comparison, charging stations, which are based on the CHAdeMO standard, are currently able to provide up to 50 kW charging power.[78] The charging standard CCS ("Combined Charging System") also currently has a maximum load capacity of 50 kW.[79]

Charging at the Tesla Supercharger is free for drivers of the Model S and Model X, i.e. the establishment and operation of the Supercharger are cross-financed by the sales proceeds of Tesla. In providing energy without a quantity-dependent payment, offered exclusively to drivers of the Model S and Model X, the Supercharger network ultimately lacks the purpose of selling energy to an indefinite number of prospective customers. Supplying customers with energy in accordance with § 3 No. 36 EnWG is therefore not the case here because none of the three variants mentioned is relevant: It is neither the generation or production of energy for supplying to customers nor sales of energy to customers and not the operation of an energy supply system.

The business model therefore looks more similar to providing energy to a hotel guest in the context of a hotel visit: In this case, the hotel is not an energy supplier or system operator, because the supply of energy is of minor importance, even if it is indispensable for the product "hotel accommodation."[80] In the case of Tesla Motors, in contrast to hotel accommodation, the Supercharger network addresses only a certain number of interested parties, because there is no general accessibility to the network, both in technical and in legal terms, since neither can other vehicles than Tesla's Model S and Model X process the high charging currents, nor is the network aimed at the general public. Therefore, the competitive element—in regard to the charging—is missing.

---

[77] Tesla Motors, Inc. (2015).

[78] CHAdeMO Association (2015).

[79] Due to the limiting at 125 amps and at 400 volts, see Federal Ministry of Economic Affairs and Energy (2015, p. 12).

[80] FCJ, case no. EnVR 68/10, recitals 14–15.

According to Art. 2 no. 19 of Directive 2009/72/EC, the supply of customers means the sale of electricity, which also does not take place here. The Supercharger network is therefore no supply system within the meaning of § 20 EnWG.

As the drivers of the Tesla Model S also do not purchase any electricity at the Supercharger, they are no final customers, see § 3 no. 25 EnWG, and Supercharger can therefore not be defined as customer facilities within the meaning of § 3 no. 24a or 24b EnWG.

References

ADAC (2013) ADAC Autotest BMW i3. https://www.adac.de/_ext/itr/tests/Autotest/AT5053_BMW_i3/BMW_i3.pdf. Accessed 07 July 2016

Bartsch M (2002) In: Bartsch M, Röhling A, Salje P, Scholz U (eds) Stromwirtschaft – Ein Praxishandbuch. Carl Heymanns Verlag, Köln

BDEW (2008) MeteringCode 2006. http://www.bdew.de/bdew.nsf/id/DE_7PDK2P_MeteringCode_2006/$file/MeteringCode_2008-05-29.pdf. Accessed 07 July 2016

Boesche KV (2013) Energiewirtschafts- und kartellrechtliche Einordnung der Ladeinfrastruktur und Auswirkungen des Ergebnisses, Unterkapitel 1 - Kriterien der Anforderungen an einen diskriminierungsfreien Zugang zur Ladeinfrastruktur. In: Boesche KV, Franz O, Fest C, Gaul AJ (eds) Berliner Handbuch zur Elektromobilität, C.H.BECK, München, pp 183–190

Boesche KV (2015) Sind Ladepunkte für Elektrofahrzeuge Letztverbraucher? RdE 2015, vol 10–11, pp 449–455

Böwing A (2013) Umsetzung der Elektromobilität im Umfeld des heutigen Elektrizitätsmarktes. In: Boesche KV, Franz O, Fest C, Gaul AJ (eds) Berliner Handbuch zur Elektromobilität. C.H. BECK, München, pp 87–100

BT-Drucksache 15/3917. Entwurf eines Zweiten Gesetzes zur Neuregelung des Energiewirtschaftsrechts

BT-Drucksache 17/6072. Gesetzentwurf der Fraktionen der CDU/CSU und FDP, Entwurf eines Gesetzes zur Neuregelung energiewirtschaftsrechtlicher Vorschriften vom 06.06.2011

CHAdeMO Association (2015) Description of the CHAdeMO standard, CHAdeMO technological strengths. Optimal output power. http://www.chademo.com/wp/technology/optimal/. Accessed 07 July 2016

de Wyl C (2015) Der Energieblog. Durchbruch für die Elektromobilität? Das Berliner Modell einer Ladeinfrastruktur. http://www.derenergieblog.de/alle-themen/verkehr/durchbruch-fuer-die-fuer-elektromobilitaet-das-berliner-modell-einer-ladinfrastruktur/. Accessed 07 July 2016

European Commission (2001) First benchmarking report on the implementation of the internal electricity and gas market 2001. https://ec.europa.eu/energy/sites/ener/files/documents/2001_report_bencmarking.pdf. Accessed 07 July 2016

European Commission (2013) Proposal for a directive of the European parliament and of the council on the deployment of alternative fuels infrastructure. COM/2013/018 final—2013/0012 (COD). http://eur-lex.europa.eu/legal-content/EN/TXT/HTML/?uri=CELEX:52013PC0018&from=EN. Accessed 07 July 2016

European Commission (2014) Press release: EU launches clean fuel strategy. http://europa.eu/rapid/press-release_IP-13-40_en.htm. Accessed 07 July 2016

Evers H-U (1983) Das Recht der Energieversorgung. 2nd edn. Nomos, Baden-Baden

Federal Cartel Office (2000) Pressemitteilung des Bundeskartellamtes vom 20.04.2000. http://www.bundeskartellamt.de/SharedDocs/Meldung/DE/Pressemitteilungen/2000/20_04_2000_Wechselgeb%C3%BChren.html. Accessed 07 July 2016

Federal Cartel Office (2000) Pressemitteilung des Bundeskartellamtes vom 25.04.2000. http://www.bundeskartellamt.de/SharedDocs/Meldung/DE/Pressemitteilungen/2000/25_04_2000_Durchleitung.html. Accessed 07 July 2016

Federal Cartel Office (2001) Pressemitteilung des Bundeskartellamtes vom 01.02.2001. http://www.bundeskartellamt.de/SharedDocs/Meldung/DE/Pressemitteilungen/2001/01_02_2001_Stromnetzbetreiber.html. Accessed 07 July 2016

Federal Cartel Office (2011) Sektoruntersuchung Stromerzeugung und –großhandel. http://www.bundeskartellamt.de/SharedDocs/Publikation/DE/Sektoruntersuchungen/SektoruntersuchungStromerzeugungStromgrosshandel-Abschlussbericht.html?nn=4143254. Accessed 07 July 2016

Federal Ministry of Economic Affairs and Energy (2015). Leitfaden zum Aufbau von Schnellladesäulen im Forschungsvorhaben SLAM. http://www.slam-projekt.de/pdfs/investoren/leitfaden.pdf. Accessed 07 July 2016

Federal Ministry for Economic Affairs and Energy (2015) Referentenentwurf des Bundesministeriums für Wirtschaft und Energie, Entwurf eines Gesetzes zur Weiterentwicklung des Strommarktes (Strommarktgesetz). http://www.bmwi.de/BMWi/Redaktion/PDF/E/entwurf-eines-gesetzes-zur-weiterentwicklung-des-strommarktes.pdf; Accessed 07 July 2016

Federal Ministry for Economic Affairs and Energy (2015) Verordnung über technische Mindestanforderungen an den sicheren und interoperablen Aufbau und Betrieb von öffentlich zugänglichen Ladepunkten für Elektromobile. http://www.bmwi.de/DE/Presse/pressemitteilungen,did=736430.html. Accessed 07 July 2016

Federal Ministry for Education and Research (2009) Nationaler Entwicklungsplan Elektromobilität. https://www.bmbf.de/files/nationaler_entwicklungsplan_elektromobilitaet.pdf. Accessed 07 July 2016

Federal Ministry of Transport and Digital Infrastructure (2009) Modellregion Hamburg. http://www.bmvi.de/SharedDocs/DE/Artikel/G/modellregion-hamburg.html?nn=36210. Accessed 07 July 2016

Federal Network Agency (2010) Case no. BK6-10-136. http://www.bundesnetzagentur.de/DE/Service-Funktionen/Beschlusskammern/1BK-Geschaeftszeichen-Datenbank/BK6-GZ/2010/2010_0001bis0999/2010_100bis199/BK6-10-136_BKV/BK6-10-136_Verpflichtung_Netzzugang.pdf. Accessed 07 July 2016

Federal Network Agency (2011) Anlage zum Beschluss BK6-06-009. http://www.bundesnetzagentur.de/DE/Service-Funktionen/Beschlusskammern/1BK-Geschaeftszeichen-Datenbank/BK6-GZ/2011/2011_0001bis0999/2011_100bis199/BK6-11-150/Konsolidierte_Lesefassung_GPKE.pdf?__blob=publicationFile&v=3. Accessed 07 July 2016

Federal Network Agency and Federal Cartel Office (2003) Bericht über die energiewirtschaftlichen und wettbewerblichen Wirkungen der Verbändevereinbarungen (Monitoring-Bericht). BT-Drucksache 15/1510

Federal Network Agency and Federal Cartel Office (2013) Monitoringbericht 2013. http://www.bundesnetzagentur.de/cln_1432/DE/Sachgebiete/Elektrizitaetund-Gas/Unternehmen_Institutionen/DatenaustauschundMonitoring/Monitoring/Monitoringberichte/Monitoring_Berichte_node.html. Accessed 07 July 2016

Federal Network Agency and Federal Cartel Office (2014) Monitoringbericht 2014. http://www.bundesnetzagentur.de/cln_1432/DE/Sachgebiete/ElektrizitaetundGas/Unternehmen_Institutionen/DatenaustauschundMonitoring/Monitoring/Monitoringberichte/Monitoring_Berichte_node.html. Accessed 07 July 2016

Feller D, de Wyl C, Missling S (2010) Ladestationen für Elektromobilität - regulierter Netzbereich oder Wettbewerb? In: ZNER 2010, pp 240–246

Fest C, Franz O, Haas G (2010) Energiewirtschaftliche und energiewirtschaftsrechtliche Fragen der Elektromobilität - Teil 1. In: ET 2010, no 4, pp 93–98

Franz O, Fest C (2013) Mögliche Markt- und Regulierungsmodelle für (öffentliche) Ladeinfrastrukturen. In: Boesche KV, Franz O, Fest C, Gaul AJ (eds) Berliner Handbuch zur Elektromobilität. C.H.BECK, München, pp 149–182

Freie und Hansestadt Hamburg (2014) Masterplan Elektromobilität Hamburg. http://www.hamburg.de/contentblob/4362700/data/pm-26-08-2014-masterplan.pdf. Accessed 07 July 2016
Hartwig M (2013) Öffentliche Ladestationen als Teil des Elektrizitätsversorgungsnetzes. In: ZNER 2013, pp 475–482
Heinlein B (2013) Energiewirtschafts- und kartellrechtliche Einordnung der Ladeinfrastruktur und Auswirkungen des Ergebnisses, Unterkapitel 2 - Einordnung der Ladesäuleninfrastruktur als Energieversorgungsnetz. In: Boesche KV, Franz O, Fest C, Gaul AJ (eds) Berliner Handbuch zur Elektromobilität. C.H.BECK, München, pp 191–200
Keil E, Schmelzer K (2010) Systemintegration von Elektromobilität, Herausforderung an das Energiewirtschaftsrecht, eine Standortbestimmung - Teil 2. In: ZNER 2010, no 5, pp 563–567
LichtBlick SE (2012) Wie E-Autos mit SchwarmEnergie® die Stromnetze stabilisieren können. http://www.lichtblick.de/privat-kunden/schwarm-energie/innovationen/schwarmmobilitaet. Accessed 07 July 2016
Löwer W (1989) Energieversorgung zwischen Staat, Gemeinde und Wirtschaft (Recht-Technik-Wirtschaft, Bd. 53). Carl Heymanns Verlag, Köln
Lüdemann V, Jürgens C, Ortmann M (2014) Datenschutz und Elektromobilität: Zum Ladevorgang an öffentlichen Ladesäulen. In: RDV 2014, no 01, pp 3–10
Mengers H (2007) Der Energiemarkt. In: Germer C, Loibl H (eds) Handbuch Energierecht. Erich-Schmidt-Verlag, Berlin
Michaels S, de Wyl C, Ringwald R (2011) Rechtsprobleme im Zusammenhang mit der Nutzung des öffentlichen Straßenraums für Elektromobilitätsanlagen. In: DÖV 2011. No. 21. pp. 831-840
Peters F (2012) Alles elektrisch?! – Chancen und Konzepte der Elektromobilität. In: Veröffentlichungen des Japanisch-Deutschen Zentrums Berlin, vol 63, pp 149–157
PricewaterhouseCoopers AG (2014) Elektromobilität: Mobiler Zählpunkt fehlt. http://blogs.pwc.de/auf-ein-watt/energiewende/elektro-mobilitaet-mobiler-zaehlpunkt-fehlt/1013/. Accessed 07 July 2016
Potacs M (2009) „Effet utile als Auslegungsgrundsatz“. In: EuR. No. 4 in 2009, pp 465–487
Reinke J (2014) Bereitstellung öffentlicher Ladeinfrastruktur für Elektrofahrzeuge: eine institutionenökonomische Analyse. Berlin
Säcker FJ (2014) Berliner Kommentar zum Energierecht, Band 1. 3rd edn. Deutscher Fachverlag GmbH. Fachmedien Recht und Wirtschaft, Frankfurt
Salje P (2006) Energiewirtschaftsgesetz – Kommentar. Carl Heymanns Verlag, Köln
Schau, Götz-Friedrich (2007) Wann ist ein Netz ein Netz? (Teil 2). In: IR 2007, vol 6, pp 122–125
Seeliger A (2000) Ist die Verbändevereinbarung zur Stromnetznutzung gescheitert? In: Wirtschaftsdienst – Zeitschrift für Wirtschaftspolitik, vol 80, no 11, pp 661–665
Senke M (2003) Elektrizitätslieferverträge nach der Liberalisierung. Humboldt-Universität, Berlin
Statista GmbH (2015) Fakten zum Thema: Wohnen in Deutschland, Analyse der Statista GmbH. http://de.statista.com/themen/51/wohnen/. Accessed 07 July 2016
Tennet Holding BV (2014) Balancing groups. http://www.tennet.eu/de/index.php?id=310&L=2. Accessed 07 July 2016
TenneT Holding BV (2014) Errichtung einer Konverterstation auf See. http://www.tennet.eu/de/netz-und-projekte/rund-um-den-netzausbau/bauphasen/netzanbindung-auf-see-offshore/errichtung-konverterstation-offshore.html. Accessed 07 July 2016
Tesla Motors, Inc. (2015) Description of the Supercharger network. https://www.teslamotors.com/supercharger?redirect=no. Accessed 07 July 2016
VDI/VDE Innovation + Technik GmbH (2012) INEES - Intelligente Netzanbindung von Elektrofahrzeugen zur Erbringung von Systemdienstleistungen. http://www.erneuerbar-mobil.de/de/projekte/foerderung-von-vorhaben-im-bereich-der-elektromobilitaet-ab-2012/kopplung-der-elektromobilitaet-an-erneuerbare-energien-und-deren-netzintegration/inees. Accessed 07 July 2016
von Hoff S (2009) Zugangsanspruch zu Elektromobilitätstankstellen. ZNER 2009:341–345
Weis E (2014) E-Mobility. Nomos, Baden-Baden

WISO, ZDF-programme (2015) „Wucher“-Tarife beim Strom-Tanken. http://www.t-online.de/auto/alternative-antriebe/id_72854296/ladestationen-preise-fuer-strom-teurer-als-verbrenner.html. Accessed 07 July 2016

Zeit Online (2015) Regierung hält an Elektroauto-Ziel fest. http://www.zeit.de/mobilitaet/2015-04/elektromobilitaet-bundesregierung-ziele-gruene. Accessed 07 July 2016

Chapter 3
Charging Stations as "*Essential Facilities*"

Electricity suppliers, in addition to the previously shown right of third party network access in accordance with § 20 para. 1 of the EnWG, may also have a right to non-discriminatory network access in accordance with competition law. This claim could arise from both Art. 102 TFEU or § 19 para. 2 no. 4 GWB. However, with § 111a EnWG a provision exists, which precludes a parallel application of § 19 para. 2 no. 4 GWB additionally to the provisions of the Energy Industry Act. An application of § 19 para. 2 no. 4 GWB could therefore only be considered if a right of network access pursuant to § 20 para. 1 EnWG were dismissed in principle.[1]

With regard to the fundamental primacy of European law, only the claim under Article 102 TFEU will therefore be examined and reads as follows:

> ***Electricity suppliers have a right to non-discriminatory access to publicly accessible charging infrastructure according to Art. 102 TFEU.***

3.1 The Benefit for the Consumer

A right to network access based on European competition law would also allow competition for the supply of electric vehicles at public charging stations and thus create the basis for new and better products. Above all, it would be possible for final consumers to charge their electric vehicle at any public charging station with only one charge card, issued by their electricity supplier.

---

[1]Whereby the dogmatic meaningfulness of § 111a EnWG must be seriously called into question, since regulatory and antitrust law have different objectives. The provision eventually results in a gap in legal protection, for example in regard to network tariffs, because even approved network tariffs may violate competition law, see ECJ, C 280/08 P *(„Telekom")*.

M. Adam, *Accelerating E-Mobility in Germany*, SpringerBriefs in Law
DOI 10.1007/978-3-319-44884-8_3

3.2 Network Access Pursuant to Art. 102 TFEU

3.2.1 Dominant Position

A right to network access pursuant to Art. 102 TFEU initially requires a dominant position within the common market or a substantial part there of. Therefore the relevant market needs to be identified and defined in terms of the relevant product market and geographic market.[2]

3.2.1.1 Product Market

Both the Federal Court of Justice[3] and the European Commission[4] use the demand-side oriented market concept[5] to define the market in material terms: This means, that all products and services that are interchangeable from the perspective of customers for property, use and price range to meet a particular need will be attributed to the relevant market.[6]

In the case at hand the service can be described as providing a publicly accessible charging point for charging an electric vehicle. Publicly accessible means in this case that the charging point is aimed at a non-predetermined, indefinite clientele and in particular no access restrictions for certain electric vehicles or their drivers exist, i.e. charging points which provide Union-wide non-discriminatory access to users.[7]

Beyond this basic definition, the question is whether there are technical differences that call for a more detailed demand-side definition. There are various connector types for example, which may separate the market for publicly accessible charging. Also, there are different charging services (normal charging and fast charging) and the ability to charge the vehicle via AC or DC power.

The various connector types should not lead to a separate product market definition for two reasons: On the one hand there are numerous adapters that allow each electric vehicle to establish a connection between the vehicle and the charging point. On the other hand it is determined by Directive 2014/94/EU that all of the normal charging points shall at least be provided with sockets or vehicle connectors of type 2 according to the standard EN62196-2 ("Mennekes plug").[8] For fast charging points, which are operated with direct current, a specification of the "CCS 2"

---

[2]European Commission (1997).

[3]FCJ, Decision on December 06, 2011, case no. KVR 95/10 („Total/OMV"), recital 27.

[4]Form CO relating to the notification of a concentration pursuant to Regulation (EEC) No 4064/89 (Annex to Regulation 3384/94), section 6.

[5]von Dietze and Janssen (2011), recital 376.

[6]FCJ, ibidem.

[7]see Art. 2 No. 7 of Directive 2014/94/EU.

[8]Annex II, No. 1.1 of Directive 2014/94/EU.

system ("Combined Charging System Combo 2") is established according to the standard EN62196-3.[9] By being able to use adapters and through the provisions of Directive 2014/94/EU a very broad interoperability is therefore ensured, so that a market definition according to different connector types is not necessary.

A technical difference, however, exists in regard to fast charging with direct current: Not all vehicles have the ability to recharge their battery via DC. The main technical difference is that charging with direct current is much faster and supplied at much higher power levels (above 22 kW). At first glance an objective definition is therefore needed, but it should be noted that any vehicle that can be charged via DC, can also be charged with AC. There is therefore a backward compatibility and to this extent a substituting possibility, but only of DC charging toward the AC charging, but not vice versa.

The shorter charging time of DC charging stations stems from the fact that the vehicle battery is always charged with direct current.[10] The electricity grid, however, at the final consumer level typically supplies AC. With conventional charging at the low voltage level a conversion of the alternating current into direct current must therefore be carried out to then charge the battery in the car. If the charging station offers DC (which means the conversion of alternating current to direct current takes place within the charging station) and the vehicle has a so-called bypass option, meaning to directly charge the battery with direct current, the battery can be charged with significantly higher performance.

A market definition for charging by means of AC and DC charging correlates to the methodology and approach of the Federal Cartel Office in regard to the current market definition on the final market for electricity.[11] The market definition by the Federal Cartel Office makes a distinction between the supply of final consumers via load profile measurement (LPM customers), via standard load profile-special contract (SLP special contract), via SLP-basic supply and via SLP heating current.[12] The Federal Cartel Office uses the respective demand behavior that ultimately manifests in technical features in the supply as a distinguishing factor. With LPM customers it is the high power and consumption intake, which is indicated by the load profile meter, see. § 12 para. 1 StromNZV. SLP customers are characterized by relatively low consumption, differentiated according to the specific legal basis of the basic supply and other contracts (special contract customers), and SLP-heating customers, which deviate most from the standard load profile (high, uniform supply at night; high consumption with lower power; depending on the temperature).

Applied to charging by AC and DC, the difference in the demand pattern stems from the fact that the vehicle has the described bypass option, so you can charge

[9]Annex II, No. 1.2 of Directive 2014/94/EU.

[10]Federal Ministry for Transport and digital Infrastructure (2014, pp. 11–12).

[11]Federal Cartel Office, Decision of November 30, 2009, case no. B8-107/09 (Integra/Thüga), recitals 24ff., and Decision of April 30, 2010, case no. B8-109/09 (RWE/SW Lingen/SW Radevormwald), recitals 26ff.

[12]Federal Cartel Office, ibidem.

directly to the battery. The charging is carried out with significantly higher power (and thus in a much shorter time) than when charging with alternating current. This is similar to LPM customers (high power intake, typically 30 kW and more)[13] and SLP customers (low power intake, typically less than 30 kW). In the future the market definition of the final consumer level could therefore be expanded by these two variants.

However, for the consideration here only AC charging stations should be dealt with, because there were only 102 charging stations with direct current in Germany according to the figures of the BDEW for the end of 2014, but 5,553 charging stations that offer AC charging.[14]

The product market therefore consists of publicly accessible charging stations that offer AC charging.

3.2.1.2 Geographic Market

The relevant geographic market comprises the area in which the undertakings concerned are involved in the supply and demand of products or services, in which the conditions of competition are sufficiently homogeneous and which can be distinguished from neighbouring areas because the conditions of competition are appreciably different in those areas.[15]

Due to the currently limited range of electric vehicles from around 100 up to 200 km one might at first be inclined to limit the relevant geographic market initially only to the urban area or the power grid area. City and regional tours can already be experienced with a single battery charge today. However, the demand is for nationwide driving with electric vehicles, so that the federal territory is the minimum size for this market from a customer perspective.

A cross-border driving will certainly also be relevant in the future, but at present the current technical limitations mean that a cross-border, electric drive takes place only in exceptional cases. These technical limitations include different types of connectors, low battery capacity and a lower density of charging stations in most neighboring European countries.

The concrete demand behavior of electricity suppliers is also decisive for the geographic market definition. Even if there should already be a concrete (and significant) demand from final consumers for cross-border charging solutions, it is most likely, however, that no electricity supplier is actually planning or even demanding cross-border network access to charging infrastructure at present. The reason for this is that the energy supplier has to maintain a balancing group in the neighboring country. The relevant national regulations in the energy sector,

---

[13]See § 2 Abs. 7 KAV.

[14]BDEW (2015).

[15]European Commission (1997), recital 8.

however, are not harmonized and the balancing markets are fragmented.[16] Therefore special know-how and a lot of effort are needed to successfully execute a cross-border supply to final consumers. The limited network capacity of the interconnectors is another obstacle. As a result, cross-border supply of electricity to final consumers will take place only rarely and if at all, then only for large industrial users in the border area between two EU member states.

As a result, the geographic market for electricity supply via charging infrastructure is therefore currently limited to German territory.

3.2.1.3 Dominant Position

According to figures of the BDEW about 5,500 public charging points were available in 2014.[17] The RWE Efficiency GmbH, a 100 % owned subsidiary of RWE AG Germany, which owns and operates over 1,500 charging points nationwide,[18] commanding a 27 % market share, already has a significant market position in Germany.

For the electricity supplier who wants to offer its own power at each of the public charging stations, however, it is not so much the question of *one* dominant undertaking, but rather the question of whether there exists a collective—i.e. oligopolistic—dominance by a variety of companies at present. The background is that a supplier of a nationwide driving product must be able to supply energy at *every* publicly accessible charging station, because otherwise a single, nationwide functioning charge card cannot be offered.

Collective Entity

The term "one or more undertakings" in Art. 102 TFEU means that a dominant position by two or more legally independent economic entities may be held, provided that they act together in economic terms as a collective entity on a given market.[19]

In particular it must be examined whether there are any economic ties between the companies concerned, which allow them, to act together independently of their competitors, their customers and consumers.[20]

---

[16]European Commission (2015, pp. 5 and 9).

[17]BDEW (2015).

[18]RWE Effizienz GmbH (2015).

[19]ECJ, Judgement of the Court of March 16, 2000, C-395/96 P („Compagnie Maritime Belge Transports"), recital 36; CFI, Judgement of the Court of January 26, 2005, T-193/02 („Piau"), recital 110.

[20]ECJ, Judgement of the Court of March 16, 2000, C-395/96 P(„Compagnie Maritime Belge Transports"), recital 42.

Currently there is not a single one charging station operator in Germany, who grants network access to third party electricity suppliers. On the contrary: The charging stations operators are connected in so-called roaming networks,[21] i.e. there are contractual agreements that allow the supply of contract customers of other charging station operators at its own charging station—but only with the energy of the respective (local) charging station operator and not just by way of third party network access.

So a case can be made that there is a collective appearance and collective performance, since a network access of third party suppliers does not take place and instead the roaming networks exclusively provide third party supply via the resale model.

Since this fact is decisive for the question of dominance, the criterion of the collective unit shall be examined more closely hereinafter. The Court of Justice states that three cumulative conditions must be met for the finding of collective dominance:

> *„first, each member of the dominant oligopoly must have the ability to know how the other members are behaving in order to monitor whether or not they are adopting the common policy; second, the situation of tacit coordination must be sustainable over time, that is to say, there must be an incentive not to depart from the common policy on the market; thirdly, the foreseeable reaction of current and future competitors, as well as of consumers, must not jeopardise the results expected from the common policy"*[22]

Market Transparency

The information of whether right of network access for third party suppliers is granted at a certain charging station or not can be considered common knowledge, since the news would spread very fast, at least within the energy sector. In that regard there is fundamental market transparency concerning the collective behavior of charging station operators. This market transparency, i.e. the knowledge of the oligopoly members of their mutual behavior, is reinforced by common umbrella organizations that largely carry out the lobbying. These umbrella organizations mainly consist of the BDEW[23] and the VKU,[24] who collectively represent almost all the incumbent energy suppliers in Germany. In its opinion on Directive 2014/94/EU (AFI Directive), the BDEW represents accordingly the position that

---

[21]E.g. http://www.ladenetz.de/ by smartlab Innovationsgesellschaft mbH or http://www.intercharge.eu/ by der Hubject GmbH.

[22]CFI, Judgement of the Court of January 26, 2005, T-193/02 („Piau"), recital 111.

[23]BDEW (2016) is the umbrella organization for the established energy and water utilities in Germany. The member companies represent around 90 % of the electricity sales in Germany, see https://www.bdew.de/internet.nsf/id/mitglieder-de.

[24]VKU (2016) represents the interests of the local energy and disposal industry in Germany, see http://www.vku.de/ueber-uns.html.

there should be no non-discriminatory network access to charging stations.[25] The VKU represents ultimately the same position in its opinion on the charging stations Regulation in which the association carries out:

> *„The expansion of the charging infrastructure is located in the non-regulated sector. A need for regulation is not clear, as there is no natural monopoly."*[26]

This means, there is a high level of market transparency between the participants of the oligopoly. This market transparency is enhanced even further by membership in the roaming network. The roaming networks provide the platform to connect the charging infrastructure—without a third party network access. These different variants of providing charging infrastructure are mutually exclusive: Once third party network access is granted, the e-roaming model loses its scope of application.

Coordination Discipline

In addition to the joint approach per se, in accordance with the cited jurisprudence of the ECJ, an incentive is also needed to encourage the participants to persevere this joint approach in the long run and not to depart from a common policy on the market.

As already shown in the preceding section, the providers of charging infrastructure are bound twofold: On the one hand via the umbrella organizations that already may exert a strong coordination discipline by themselves. Secondly, the charging station operators are connected over the roaming networks, which provide a maximum incentive with the concrete contractual provisions for roaming to sustain the joint approach.

The treaties of the e-roaming networks ultimately manifest the coordinated behavior.

Robustness Against the Competition

The last requirement in this joint approach is a certain robustness of the collective cooperation in regard to possible competitors or consumer behavior. Thus, the collective entity unit must in principle be able to hold its position against competitive aspirations.

This can be affirmed as well in the present case. Generally, the robustness of the system is evident because neither are the roaming models questioned as such nor is third party network access to charging stations seriously considered or discussed in

---

[25]See BDEW (2013), no. 3.2.

[26]VKU (2015), p. 4.

the industry. This circumstance is therefore somewhat surprising as in the telecommunications sector the roaming model has a bad reputation as a price driver for phone calls and mobile surfing abroad[27] and therefore currently the abolition of roaming charges are being addressed specifically in the telecommunications sector.[28]

The market power of the oligopoly is very clear in the case of Hamburg: In the area of the distribution system of Hamburg non-discriminatory network access to public charging stations was implemented as part of a project funded with public money, see Sect. 2.3.5. However, instead of continuing this successful model and to campaign for a nationwide application, the City of Hamburg ended this project in 2014 and is now also about to implement a roaming model.[29]

The coordination discipline of the charging infrastructure operators has thus paid off: The publicly accessible charging points in Hamburg are no longer supplied via network access, but rather have used the roaming business model since mid-2015 and thus are now also part of the oligopoly. The pressure of the oligopoly was in fact such that until the end of 2014 no roaming or supplying via network access on public charging stations by energy suppliers from abroad took place in Hamburg, because charging station operators from other cities refused to cooperate with Stromnetz Hamburg GmbH. The result was that motorists from outside were only able to use the charging stations in Hamburg, if a contract for a charge card with Vattenfall or Hamburg Energy GmbH, the only two suppliers in Hamburg in 2014, was concluded. This was not attractive at all for motorists from outside of Hamburg and led de facto to the isolation of Hamburg's charging infrastructure.

For competitors, meaning third-party suppliers that have no charging infrastructure, only the possibility to participate through a roaming agreement therefore remains at present. But since this agreement cannot ensure without further ado that green electricity is supplied all the time, so far a majority of independent green electricity providers in Germany have declined to participate in such a roaming solution. This is another proof that the oligopoly of the charging station operators knows how to stand up to competitors.

---

[27]The European Commission (2016) clearly states in regard to roaming charges:

> „*…they are a market distortion with no rational place in a single market – they teach users to fear their phones instead of using them.*" http://ec.europa.eu/digital-agenda/en/roaming.

[28]Thus explicitly declared in No. 1 of the priorities of Jean-Claude Juncker (2015), President of the European Commission, http://juncker.epp.eu/node/151.

[29]This was confirmed by the responsible staff member of Stromnetz Hamburg during an expert talk with the author in Hamburg on January 26, 2015. The website of Stromnetz Hamburg GmbH euphemistically describes the return to the roaming model in the section „innovations for the grid of the future" as „new access option" for charging stations, http://www.stromnetz.hamburg/ueber-uns/innovationen/.

3.2.2 *Abusive Conduct*

The central question now is whether the refusal of the charging station operator to grant third party suppliers network access to the charging stations can be deemed as abusive behavior at all.

This would be the case if the publicly accessible charging station must be defined as an essential infrastructure within the meaning of the "*essential facilities*"-doctrine. The "*essential facility*"-doctrine has its roots in the US, where this doctrine is, interestingly enough, most controversial.[30] In European case law, the specific requirements of the doctrine were introduced with the ECJ decision "*Bronner*"[31] and applied within Art. 102 TFEU (formerly Art. 82 EC). After "Bronner" the essential facility must not only be essential, it must be indispensable.[32] According to the statements of the ECJ in the decision "IMS Health" an essential facility exists, where the following conditions are fulfilled:

- the undertaking which requested the access to the essential facility intends to offer, on the market in question, new products or services not offered by the owner of the essential facility and for which there is a potential consumer demand;
- the refusal is not justified by objective considerations;
- the refusal is such as to reserve the market for the products to the owner of the essential facility in the Member State concerned by eliminating all competition on that market.[33]

These three conditions are to be subsequently checked for the market of public charging stations.

3.2.2.1 New Product

According to the requirements of the ECJ, it must be a new product that is not yet available and for which there is a potential demand.[34]

In the present case the product in question is a nationwide supply of electric vehicles with electricity by means of publicly accessible charging stations—with just a *single* charge card. The price of the charging is set by the supplier and, in particular, this will most likely be a nationwide uniform price. The customer would then get *efficient, nationwide access* to charging stations with a *transparent, uniform price*.

---

30 Pitofski et al. (2002), pp. 443–448.

31 ECJ, Judgement of the Court of November 26, 1998, C-7/97.

32 ECJ, ibidem, recitals 45f.

33 ECJ, Judgement of the Court of April 29, 2004, C-418/01 („IMS Health/NDC Health"), recital 52.

34 ECJ, ibidem, recital 49.

Such a product does not yet exist, although the roaming networks basically work towards a similar product. In particular, there is no universal coverage with only one charge card. And the price always varies from charging station to charging station, even if the charge card is part of a roaming network.

The uncomplicated charging of a vehicle is considered to be a central point for the success of electric mobility, so that the final consumer will have a great interest in such a product. Thus, there is a potential demand for this product.

No Otherwise Access

The infrastructure must also be indispensable for the implementation of such a product. This raises in particular the question whether there are other ways to implement this product, or if the infrastructure facility cannot also be built by the third party itself.

The roaming networks aim, by their own account, to enable a nationwide network of charging stations and to provide the costumers optimal coverage in at least the medium to long term. Even the problem of the lack of price transparency in some places, could certainly be solved.

However, the roaming model causes systemic market foreclosure on the upstream market of wholesale of electricity and furthermore carries the risk of a margin squeeze in the downstream market of customer supply. Thus, the participation fee for an electricity supplier in a roaming network ranges from 20 to 40 EUR per year and customer, plus other costs, such as additional maintenance fees.[35] This amount represents almost exactly the margin that can be achieved when supplying household customers, as a glance at the numbers of the Monitoring Report of the Federal Network Agency and the Federal Cartel Office shows:

According to the report the share of the retail electricity price "*energy procurement, sales, other expenses and margins*" accounted for 6.67 cents/kWh in case of an annual consumption of 3,500 kWh in 2013.[36] Subtracting 3.78 Cent/kWh[37] for the procurement of electricity, that leaves 2.89 Cent/kWh for the gross margin, i.e. in case of an annual consumption of 3,500 kWh that adds up to 10,115 EUR per customer. According to an analysis by the consulting firm LBD, in 2006 selling expenses amounted to 53 EUR plus the cost for the initial acquisition of the customer.[38] Selling expenses were thus in a range of 60 to 70 EUR (if not even more).

---

[35]The author has received contract documents and pricing conditions for 2013 for participation in a certain roaming network.

[36]Federal Network Agency and Federal Cartel Office (2015), p. 164.

[37]Statista GmbH (2015), Average spot market price at power broker EEX, http://de.statista.com/statistik/daten/studie/154012/umfrage/grosshandelspreise-fuer-strom-in-deutschland-seit-2008/. In fact, the electricity procurement will have been more expensive, because the electricity is procured in a structured manner, i.e. at different times, and costs for balancing group management arise in addition to the procurement costs.

[38]bne Association of Energy Market Innovators (2006), pp. 24–25.

The remaining margin of approximately 30 to 40 EUR per customer is thus the profit of the electricity supplier. This profit is now declining to zero, however, due to the additional cost of roaming. No wonder, because the use of infrastructure is paid for twice: On the one hand by the network tariff and, secondly, by the roaming fee. Thus the roaming models are no serious alternative to third party network access.

No Duplicability

There remains the question of duplicability: Why should an energy supplier not build itself charging stations?

This involves the petrol station argument that says: Petrol stations are not "essential facilities", why should it be different with charging stations?

It is obvious that by no means all the charging stations have been built, which would be necessary for a nationwide driving experience in Germany. The electricity supplier that demands network access to the charging stations of other providers could perhaps be referred to build its own charging stations and operate its own network.

First, the comparison with the stations is flawed in so far as that it is electricity, which is distributed via charging stations and is a exclusively grid-bound product,[39] while gasoline or diesel are liquid goods, that are not transported via pipes, but by tank trucks. This ultimately leads to a classification of the publicly accessible charging station as part of the energy supply network, as described and discussed in Sect. 2.3.

But even without recourse to the legal classification the comparison with a conventional petrol station is flawed, because the most obvious difference from a customer perspective is the payment method: It may be assumed as certain that for petrol or diesel no contracts are offered with a 12 month term and a monthly standing charge, in other words: Refuelling with conventional fuel is never a continuing obligation, but a one-time transaction. With charging stations, however, the continuing obligation dominates as a product or to put it differently: The energy suppliers, which operate the charging stations, behave like electric utilities and not like a petrol station operator. Provided that all charging stations accessible to the public were accessible with conventional means of payment, i.e. cash, debit card and credit cards, this possibly would create a situation similar to that at the gas stations and competition between the charging station operators would in principle be conceivable.[40] But at present it's a completely different product, namely a power supply contract, and thus it's a very different market and not the herein discussed relevant market. A market structure, such as the one to be found with gas stations,

[39]Federal Network Agency (2014), p. 12.

[40]However, in this case there would certainly be the risk that this competition would look just like the market situation of petrol stations today—hardly a convincing idea.

does not exist due to the lack of payment methods offered for publicly accessible charging stations.

However, there is a further essential difference between the operation of a petrol station and the operation of a charging station, which is ultimately the reason for the different classification.

Just like the power lines and other grid facilities as well, the majority of publicly accessible charging stations is also located in the public sector, i.e. on public land. This can be regulated directly via a provision in the concession contract to the electricity grid[41] or via separate concession contracts, e.g. in Hamburg,[42] Berlin[43] or Aalen.[44] These concessions are almost always designed as an exclusive right, with the exception of research projects, which have sometimes been awarded in parallel to different infrastructure providers; however, those projects represent an inherently low number of users and are usually limited to a short duration. Since the spaces for public charging stations must always be connected to a parking lot, these areas are scarce especially in urban areas. The competition for these concessions is in full swing[45] and the tender documents for the public charging infrastructure in Berlin show that these are large-scale infrastructure projects with many interested parties.[46]

As a result, the duplicability must thus already be ruled out because the public land will be awarded exclusively via concessions and the parking spaces are very limited in the public domain in fact and for legal reasons. A competitor therefore lacks the possibility to build competing charging stations in sufficient quantities on public property.

That leaves the question of whether it is possible to build charging stations on private ground which are open to the public and thus enable a parallel charging infrastructure and could thus create a competitive alternative to existing public charging stations. To enable a nationwide supply of electric vehicles, thus being able to offer a charge card that allows you to charge an electric vehicle nationwide, the supplier must build a very dense network of charging stations, which can not be achieved economically by a single supplier. It is not enough to build a few charging

---

[41] GAR Baden-Württemberg (2010), § 10.

[42] See Stromnetz Hamburg GmbH (2015), § 16 Konzessionsvertrag in conjuction with no. 5.2.12 and Appendix 3 (Bericht zur Entwicklung der Ladeinfrastruktur) of „Kooperationsvereinbarung zwischen der Stadt Hamburg und Stromnetz Hamburg GmbH"; https://www.stromnetz.hamburg/ueber-uns/auftrag/.

[43] Berliner Senatsverwaltung für Stadtentwicklung und Umwelt (2015).

[44] Stadt Aalen (2011).

[45] Some interesting and well worth reading remarks that tell how RWE acquired the concessions in Berlin: Technische Universität Berlin (2011, pp. 16-22).

[46] Tendering procedure of the City of Berlin of October 25, 2012: „Lieferung, Errichtung und Betrieb von öffentlich zugänglichen Ladeeinrichtungen für Elektrofahrzeuge", tenders electronic daily (2012) (Supplement zum Amtsblatt der Europäischen Union); http://ted.europa.eu/udl?uri=TED:NOTICE:339170-2012:TEXT:DE:HTML&src=0.

stations, it would have to be thousands[47]—per supplier mind you! Thus the situation does not present itself differently than with other infrastructures, such as energy, telecommunications and rail networks: A duplicity is technically feasible, but almost never economically viable.

Furthermore, rest areas available on motorways and main roads are located on public ground and hardly on private grounds at all. Charging stations for long distance travel therefore lack sufficient alternatives on private property.

Last but not least, public charging stations are infrastructure facilities, which are characterized by high fixed costs and falling average costs.[48] These fixed costs, the costs for the establishment of a charging station network, would also be irreversible costs, i.e. a competitor must fear high losses at the exit from this market segment. Regarding publicly accessible charging infrastructure a strict subadditivity can be observed—just like the original power grid -, i.e. a single company can produce the entire quantity of goods cheaper than several smaller companies combined. A competition between charging station operators is therefore not economically meaningful and cannot be expected.

Even if an energy supplier should try the construction of its own network, which would as described make little sense economically, this does not solve the existing challenge of achieving nationwide mobility through a dense network of charging stations. Without involving the existing charging stations, the optimal benefits could not be achieved because there would still be gaps in the supply of electric vehicles by publicly accessible charging stations.

After all, it must be noted that a duplicability of the network does not exist. Each publicly accessible charging station is therefore indispensable within the meaning of the "essential facilities" doctrine.

Refusal Without Justification

The refusal to grant network access, however, could be objectively justified.

Here the argument is continually invoked by the charging stations operators that the system of network access to the charging stations would be associated with significant expenses that would be disproportionate to the achievable benefits.[49] As already shown in Sect. 2.3.5 with reference to the Hamburg model, granting network access is possible without any problems and especially achievable with existing IT processes. The adjustments that need to be made to the IT system, therefore, are part of the conventional operation. Granting the right to network access cannot therefore be considered unreasonable or disproportionate.

Another argument that is often invoked is the "death" of existing business models and the risk that current providers could drop out of the market, if they were compelled

[47]For example, Aral operates 2377 petrol stations in Germany, see Statista GmbH (2016).

[48]See also Reinke (2014), p. 144; rejecting: Franz (2011, p. 16).

[49]BDEW (2013), no. 3.2.

to grant network access to their charging stations.[50] First, it should be noted that the overwhelming number of charging infrastructure operators are incumbent energy suppliers.[51] For these companies, the implementation of network access is just no hurdle or undue burden, but is a conventional business operation. However, there are occasional local vendors who actually conduct their main business in other sectors, such as bakeries that operate a public charging station on the side. In those cases a closer look is warranted: If it really is a publicly accessible charging station and the recharging of electric vehicles also pursues a commercial purpose (i.e. a payment is made in exchange for charging), a right to network access must be granted in principle. If a company behaves like an infrastructure company, it must be measured by the same standards as any other infrastructure company in this sector. In most cases, however, there will be no public access to the charging station or there is a lack of an immediate commercial purpose, i.e. there is no sale of electricity.

These arguments are therefore not convincing: Existing business models will not be destroyed and no players will be pushed out of the market because they are obliged to grant network access.

3.3 Capability of Affecting Trade Between Member States

The refusal to grant network access at charging stations concerns the entire country. Thus competition is eliminated within the largest EU Member State. This also applies to cross-border trade, as far as a potential supplier from abroad wants to offer its electricity at German charging stations. The refusal to grant network access at every public charging station in Germany is therefore capable of affecting trade between Member States in the Union.[52]

3.4 Result

Electricity suppliers have a right to network access at public charging stations in accordance with Art. 102 TFEU. These charging stations are essential infrastructure facilities within the meaning of the "essential facilities" doctrine.

---

[50]VKU (2015), p. 7; BDEW, ibidem.

[51]See Annex A—Overview: Operators of Charging Stations.

[52]This assessment is also in line with the position of the European Commission (2004), recital 78:

> *„Horizontal cartels covering the whole of a Member State are normally capable of affecting trade between Member States. The Community Courts have held in a number of cases that agreements extending over the whole territory of a Member State by their very nature have the effect of reinforcing the partitioning of markets on a national basis by hindering the economic penetration which the Treaty is designed to bring about."*

In accordance with § 22 para. 3 GWB in addition to the claim from Art. 102 TFEU, a right to network access in accordance with § 19 para. 2 no. 4 GWB can be considered as well. However, that would require in accordance with § 111a EnWG that the application of the EnWG is excluded in this case (contrary to the statements contained in Sect. 3.2) or that the provisions of § 111 EnWG are considered to be incompatible with European Law and thus must be ruled ineffective.[53]

References

BDEW (2013) BDEW-Stellungnahme zum Vorschlag der Europäischen Kommission für eine Richtlinie des Europäischen Parlaments und des Rates über den Aufbau der Infrastruktur für alternative Kraftstoffe. https://www.bdew.de/internet.nsf/id/20130426-bdew-stellungnahme-zum-vorschlag-der-europaeischen-kommission-fuer-eine-richtlinie-des-euro?open&ccm=300110010050. Accessed 07 Jul 2016

BDEW (2015) Ausbau der Ladeinfrastruktur für Elektroautos in Deutschland auf gutem Weg (3. figure). http://www.voltimum.de/artikel/ausbau-der-ladeinfrastruktur-fuer-elektroautos-deutschland-auf-gutem-weg. Accessed 07 Jul 2016

BDEW (2016) BDEW-Mitglieder und Ehrenmitglieder. https://www.bdew.de/internet.nsf/id/mitglieder-de. Accessed 07 Jul 2016

Berliner Senatsverwaltung für Stadtentwicklung und Umwelt (2015). Press Release on 09 January 2015. http://www.stadtentwicklung.berlin.de/aktuell/presse-box/archiv_volltext.shtml?arch_1501/nachricht5465.html. Accessed 07 Jul 2016

BNE Association of Energy Market Innovators (2006) BTOElt-Gutachten—Wettbewerbliche Wirkung und rechtliche Zulässigkeit der BTOElt im liberalisierten Strommarkt. http://www.neue-energieanbieter.de/en/system/files/06_10_31_btoelt_gutachten.pdf. Accessed 07 Jul 2016

European Commission (1997) Commission notice on the definition of relevant market for the purposes of community competition law (97/C 372/03). http://eur-lex.europa.eu/legal-content/EN/TXT/HTML/?uri=CELEX:31997Y1209(01)&from=EN. Accessed 07 Jul 2016

European Commission (2004) Commission notice—guidelines on the effect on trade concept contained in articles 81 and 82 of the Treaty. 2004/C 101/07

European Commission (2015) Consultation on a new energy market design. http://ec.europa.eu/energy/sites/ener/files/documents/1_EN_ACT_part1_v11.pdf. Accessed 07 Jul 2016

European Commission (2016) Digital agenda for Europe—a Europe 2020 initiative. http://ec.europa.eu/digital-agenda/en/roaming. Accessed 07 Jul 2016

Federal Ministry for Transport and Digital Infrastructure (2014) Öffentliche Ladeinfrastruktur für Städte, Kommunen und Versorger. https://www.now-gmbh.de/de/service/publikationen. Accessed 07 Jul 2016

Federal Network Agency (2014) Decision of May 20, 2014. Case no. BK7-13-073. http://www.bundesnetzagentur.de/DE/Service-Funktionen/Beschlusskammern/1BK-Geschaeftszeichen-Datenbank/BK7-GZ/2013/2013_001bis099/BK7-13-073_BKV/Beschluss.pdf. Accessed 07 Jul 2016

Federal Network Agency and Federal Cartel Office (2015) Monitoringbericht 2014. http://www.bundesnetzagentur.de/cln_1421/DE/Sachgebiete/ElektrizitaetundGas/Unternehmen_Institutionen/DatenaustauschundMonitoring/Monitoring/Monitoringberichte/Monitoring_Berichte.html?nn=266276. Accessed 07 Jul 2016

[53]This is the opinion of the author; opposing view: Higher Reginal Court Munich, Judgement of May 20, 2010, case no. U (K) 4653/09, recital 42.

Franz, O. (2011). Elektromobilität zwischen Wettbewerb und Regulierung—Zur Regulierungsnotwendigkeit von (öffentlicher) Ladeinfrastruktur. www.wik.org/fileadmin/Konferenzbeitraege/netconomica/2011/Franz.pdf. Accessed 07 Jul 2016

GAR Baden-Württemberg (2010). Kommunalfreundlicher Muster-Konzessionsvertrag. http://www.gar-bw.de/fileadmin/gar/pdf/Energie_und_Klima/kommunalfreundlich-er_Musterkonzessionsvertrag_der_Gruenen_-_Fassung_Maerz_2010.pdf. Accessed 07 Jul 2016

Juncker, JC (2015) My priorities. http://juncker.epp.eu/node/151. Accessed 07 Jul 2016

Pitofski R, Patterson D. Hooks J (2002) The essential facilities doctrine under U.S. Antitrust Law. Antitrust Law J 2002 70:443–462. http://scholarship.law.georgetown.edu/facpub/346/. Accessed 07 Jul 2016

Reinke J (2014) Bereitstellung öffentlicher Ladeinfrastruktur für Elektrofahrzeuge. Berlin. https://depositonce.tu-berlin.de/handle/11303/4462. Accessed 07 Jul 2016

RWE Effizienz GmbH (2015) Press release of April 22, 2015. http://www.rwe.com/web/cms/de/250036/rwe-effizienz-gmbh/presse-news/pressemeldung/?pmid=4012918. Accessed 07 Jul 2016

Stadt Aalen (2011) Press Release on 19 April 2011. http://www.aalen.de/elektromobilitaet-in-aalen–konzessionsvertrag-unterschrieben.26647.25.htm. Accessed 07 Jul 2016

Statista GmbH (2015) Average spot market price at power broker EEX. http://de.statista.com/statistik/daten/studie/154012/umfrage/grosshandelspreise-fuer-strom-in-deutschland-seit-2008/. Accessed 07 Jul 2016

Statista GmbH (2016) Anzahl der Aral-Tankstellen in Deutschland in den Jahren 2003 bis 2015. http://de.statista.com/statistik/daten/studie/155441/umfrage/anzahl-der-aral-tankstellen-in-deutschland/. Accessed 07 Jul 2016

Stromnetz Hamburg GmbH (2015) Innovationen für die Netze der Zukunft. http://www.stromnetz.hamburg/ueber-uns/innovationen/. Accessed 07 Jul 2016

Stromnetz Hamburg GmbH, Kooperationsvereinbarung zwischen der Stadt Hamburg und Stromnetz Hamburg GmbH. https://www.stromnetz.hamburg/ueber-uns/auftrag/. Accessed 07 Jul 2016

Technische Universität Berlin (2011) Die politischen Rahmenbedingungen der Genehmigung von Ladeinfrastruktur für Elektroverkehr im öffentlichen Raum. http://www.ivp.tu-berlin.de/fileadmin/fg93/Forschung/Projekte/e-mobility/Ergebnisbericht_Policy-Analyse__2_.pdf. Accessed 07 Jul 2016

TED Tenders Electronic Daily (2012) Berlin—Auftragsbekanntmachnung: Lieferung, Errichtung und Betrieb von öffentlich zugänglichen Ladeeinrichtungen für Elektrofahrzeuge. http://ted.europa.eu/udl?uri=TED:NOTICE:339170-2012:TEXT:DE:HTML&src=0. Accessed 07 Jul 2016

VKU (2015) VKU-Stellungnahme zum Entwurf der Ladesäulenverordnung. http://www.vku.de/energie/unternehmensstrategien/elektromobilitaet/vku-stellungnahme-zum-entwurf-der-ladesaeulenverordnung.html. Accessed 07 Jul 2016

VKU (2016) Wir über uns—VKU: der Spitzenverband der kommunalen Wirtschaft. http://www.vku.de/ueber-uns.html. Accessed 07 Jul 2016

von Dietze P, Janssen H (2011) Kartellrecht in der anwaltlichen Praxis, 4th edn. C.H.BECK, München

Chapter 4
Price Abuse Due to Market Dominance

As previously shown, there is oligopolistic behavior by the operators of the charging stations. It seems reasonable to suppose that there is not only an abuse of the dominant position in the area of access to the infrastructure but also an abusive behavior in pricing. Specifically, it is a suspicion of price abuse, i.e. unreasonable prices are asked for charging vehicles at public charging stations.

The claim to be examined reads as follows:

In the area of publicly accessible charging infrastructure cases of price abuse in accordance with Article 102 TFEU are observed.

4.1 Benefit for the Consumer

If an abusive behavior can be detected in the form of price abuse, intervention by the competition authorities would be required. Reasonable prices for the end user would be the consequence.

4.2 Price Abuse Pursuant to Art. 102 TFEU

4.2.1 Relevant Market

As already explained in Sect. 3.2.1.1, the relevant market in this present case is a nationwide market for the operation of charging infrastructure with AC charging. More precisely, the relevant market consists of charging stations that offer AC charging with a charging power of less than 22 kW.

But unlike in the previous section, hereinafter the effect on the downstream market for retail supply is to be reviewed. It is a final customer market, where electricity products for the charging of electric vehicles are offered by the operators of public charging stations.

M. Adam, *Accelerating E-Mobility in Germany*, SpringerBriefs in Law
DOI 10.1007/978-3-319-44884-8_4

This is a nationwide market as well and the considerations in Sect. 3.2.1.2 regarding the geographic market definition can be referenced.

4.2.2 Dominant Position

The oligopolistic structure initially described in Sect. 3.2.1.3 can also be assumed in regard to possible price abuse. The question is whether the collective entity has the three required properties of market transparency, coordination discipline and robustness against potential competitors or consumer behavior also in regard to the pricing in the downstream market of retail supply.

In principle, the final consumer can select a provider of his choice and will also be able to charge at numerous other charging stations of other utilities because the utility selected by the client will have one or more roaming agreements concluded with other operators of charging stations. Since the customers therefore basically have a choice, competition should actually arise between operators of charging stations and therefore pressure on prices should be noticable.

The roaming agreements provide, however, that the customer must pay the price demanded by the respective operator of that charging station.[1] The terms of STAWAG[2] for the charge card say for example, in § 5, "roaming":

> "*The charging at the charging infrastructure of roaming partners is always subject to the terms and conditions of roaming partners.*"[3]

The result of these contractual structures is that the operator of the charging infrastructure can set his prices independently of the offers of its roaming partners. The roaming agreement therefore ultimately provides consumers with the availability of many charging stations of different operators, however, the roaming agreement allows for each charging station operators to set the price independently and to customize the terms of use for its own charging stations. The roaming agreement thus acts like a demarcation agreement and thus stands in historical tradition with the electricity and gas supply before 1998 in Germany, as shown in Sect. 2.2.1 of this study. The prices are exclusively set by the operator of the charging infrastructure. This means, there is no competition *at* the charging station by third parties.

The charging stations operators therefore act as a collective entity regarding the downstream market for final consumer supply, because the roaming agreements act

---

[1]E.g. The New Motion GmbH (2016), see section „Ladetarife“, http://www.thenewmotion.de/produkte/ladekarte/; Plugsurfing GmbH (2016), see „FAQ—Wie wird der Aufladepreis bestimmt?“, https://www.plugsurfing.com/de/faq.

[2]Stadtwerke Aachen AG, partner of the roaming network „Ladenetz“.

[3]Stadtwerke Aachen AG (2015), http://www.stawag-emobil.de/fileadmin/emobilitaet/content/Dokumente/Antrag_und_AGB_Ladekarte.pdf.

as a non-aggression pact. Each charging station operator has a dominant position in the market. This dominance is to be explained in somewhat more detail:

The local demarcation of the areas of activity of the respective charging stations operators is nearly congruent with the geographical boundaries of the distribution system, because the charging stations are operated most of the time by the local energy utiltity and the respective distribution system is operated by the subsidiary of the local energy utility. In most cases, however, the public utility companies (municipal utilities) are exempted pursuant to § 7 para. 2 EnWG from the unbundling provisions, i.e. distribution, power and, indeed, the operation of the charging stations are integrated and therefore operated by one and the same company.

One exception to this rule that the charging station is eventually operated by the company that is either responsible for the distribution system or is connected under corporate law with the local grid operator, is where, due to publicly funded projects, a small number of third-party charging stations have been built as well. Another exception is the establishment of publicly accessible charging stations on private ground (so-called semi-public charging stations).

An analysis of data from http://de.chargemap.com for the urban area of Hamburg shows the distribution provided by Table 4.1.[4]

Stromnetz Hamburg GmbH therefore operates 62 % of publicly accessible charging stations in the greater Hamburg area and has a paramount market position, while the other operators follow only at a considerable distance. The Herfindahl-Hirschmann-Index[5] shows a correspondingly high market concentration in this area with a value of 4,454.

This allocation between the local distribution system operators (or the energy supplier legally affiliated with the local distribution system operators) and other suppliers can ultimately be found in all distribution system areas in Germany, because currently there are nationwide only three significant attackers in this market: The New Motion, ubitricity and RWE.

The New Motion has thereby chosen the path of so-called semi-public charging stations, which means installing publicly accessible charging stations on private grounds.

The New Motion states on its website that users have "*currently access to more than 550 charging stations by The New Motion and our roaming partners in Germany*".[6] Ladenetz, the biggest roaming network partner, in turn refers to the number of its own charging points of around 450,[7] so that the number of nationwide

[4]The data was retrieved 04 July 2015.

[5]European Commission (2004), recital 20, http://eur-lex.europa.eu/legal-content/EN/TXT/HTML/?uri=CELEX:52004XC0205%2802%29&from=DE.

[6]The New Motion GmbH (2015), http://www.thenewmotion.com/de/kontakt/fragen-und-antworten/.

[7]Smartlab Innovationsgesellschaft mbH (2015), http://www.ladenetz.de/laden.

Table 4.1 Operators of public charging stations in Hamburg

Charging station operator	Number	Annotations
Stromnetz Hamburg GmbH	62	Distribution system operator Hamburg
RWE-Konzern	12	Private grounds
Other suppliers	21	E.g. car dealerships, car parks, hotels
The New Motion	5	Private grounds
Aggregate	100	

charging stations owned and operated by The New Motion should amount to about 100.

Ubitricity on the other hand, is a start-up that offers an innovative charging solution that is not based on the conventional charging infrastructure, but a mobile charging solution composed of a charger cable with an integrated counter and cost-effective charging points that can be installed as simple outlets, even at streetlights.[8] To date, however, there are no figures for ubitricity regarding the number of publicly accessible charging points. The company rather seems to be still in the start and project phase.[9] A pilot project in Hamburg, where streetlights should have been upgraded with publicly accessible charging points by ubitricity, was rejected.[10]

The expansion of RWE charging stations has taken place mainly within the grid areas of the RWE Group. As evidenced by the website of RWE Efficiency GmbH there are some charging stations that have not been built inside the grid areas of the RWE Group, such as the 22 charging stations in northern Germany, which have been built on private grounds.[11] But with the exception of 104 charging stations in the Berlin area, which date back to a publicly funded pilot project in 2009 called "eMobility ICT",[12] the remaining metropolitan areas of charging stations have been built in the grid areas of the RWE Group. A significant competitive pressure is therefore not generated by the RWE charging stations.

Because of the barriers described in Chap. 3 (no duplicability, exclusive award of concessions, high fixed costs) lack of competition between operators charging stations is to be observed.

It is thus that the oligopoly of the local distribution system operators or the respective energy supply company legally affiliated with the DSO holds the undisputed dominant position in the market of charging infrastructure operators.

[8]The concrete charging solutions are shown at the webseite of Ubitricity under „Lösungen", https://ubitricity.com/de/l%C3%B6sungen.

[9]Ubitricity (2015), https://ubitricity.com/de/support/faqs.

[10]Hamburger Abendblatt (2015), http://www.abendblatt.de/hamburg/article205337527/Laternen-als-Zapfsaeule-Hamburg-zoegert.html.

[11]RWE Effizienz GmbH (2016), https://www.rwe-mobility.com/web/cms/de/1195202/emobility/rwe-ladesaeulenfinder/.

[12]Technische Universität Berlin (2011), http://www.ivp.tu-berlin.de/fileadmin/fg93/Forschung/Projekte/e-mobility/Ergebnisbericht_Policy-Analyse__2_.pdf.

4.2.3 *Abuse of the Dominant Market Position*

Again, more than ever, the principle applies that a dominant position in itself does not provide a sign of abusive behavior. In the present case the oligopoly does not necessarily lead to a widespread price increase, on the contrary: There are many suppliers that still offer charging at their charging stations complimentarily. Therefore a detailed examination of the respective suppliers and their their offered charging products at the charging stations is required.

4.2.3.1 Abusive Pricing

Pursuant to Article 102 para. 2 lit. a TFEU the fact that a company directly or indirectly imposes unfair prices may be an abuse. The imposition of excessive prices is probably the "classic" among the abuses due to market power, although the European Commission is rather cautious in sanctioning excessive prices.[13]

In the jurisprudence of the ECJ both the comparable market concept[14] as well as the price-cost analysis[15] are recognized as tools to determine any undue price increases.

Comparable Market Concept

A comparison between rates for e-mobility by various companies is to compare objectively the same products, i.e. applying the so-called factual comparative market concept.[16] However, the comparison of products on the same market can be unrewarding if there is already a suspicion of the existence of an oligopoly, because the risk exists that unfair pricing structures remain undetected.[17]

In addition to the comparison of the rates of charging station operators, a comparison with other electricity products may come into question, as far as a similarity of the markets is given. The demand of end consumers in the market for electricity products at charging stations can be compared very well with the demand of domestic and commercial customers within the meaning of § 3 no. 22 of the Energy Act, i.e. household or industrial and agricultural customers with a consumption up to 10,000 kWh per year. These small customers are also typically connected to the low voltage system. These key figures also apply for a customer at a public charging station. Except for the technical means by which the electricity is supplied—the house connection line on the one side and the charging station on the other side, these products are therefore very similar.

---

[13]Eilmansberger and Bien (2015), recital 201.

[14]Fuchs and Möschel (2012), recitals 180–185.

[15]Schröter and Bartl (2015), recital 192; Terhechte (2008), recital 3.97.

[16]Bulst (2014), recitals 168–172.

[17]FCJ, Judgement of June 28, 2005, case no. KVR 17/04, p. 12.

This comparison is also useful because the market of conventional energy supplies for household customers is at least served by several competitors. Accordingly, a query on a price comparison portal shows more than 20 tariffs for the same consumption case.[18] Thus, there is a reasonable assurance that the household customer tariffs do not contain an anti-competitive price increase.

To verify a potential excessive price, charging tariffs should therefore be compared with each other and also with household tariffs.

Subsequently, the twelve largest operators of publicly accessible charging stations are shown and an annual consumption of 1,000 kWh was thereby assumed (see Table 4.2).

Some of the suppliers offer a charging product with a contract, some offer also charging products without a contract, e.g. Pre-paid charge cards or direct payment options, e.g. credit card, PayPal or SMS. In so far as time-based tariffs (EUR/min) are offered, these tariffs have been converted into amount-based prices (EUR/kWh) on the basis of charging with a BMW i3 for reasons of comparability.

It turns out that price increases of up to 22 % can be reported for products with contract commitment compared to an average household electricity tariff.

Comparing the charging products without a contract, at least in the case of RWE and EnBW a substantial price increase of over 300 % can be identified.

Price-Cost-Analysis

As already mentioned, the comparable market concept reaches its limits when the overall market has an inflated price level, as is typical for monopoly and oligopoly markets. Here the price-cost analysis can provide valuable information as far as an examination of the costs is possible. In the area of electricity supply to final consumers fortunately a very extensive cost transparency is given so that here the price-cost analysis is a suitable instrument.

The price-cost analysis can also provide knowledge about the extent to which the surcharges for domestic electricity tariffs, verified partly in the previous section, have an objective justification, for example, in the form of additional costs.

The costs of consumption of 1,000 kWh/a can be broken down as shown in Table 4.3.

This cost report is based on the price list, approved by the Federal Network Agency, of BW Netze GmbH, the distribution system operator of EnBW AG, with validity from 01.01.2015,[19] as well as self-calculated costs of the charging station and the built-in meter of the charging station.[20] On this basis the annual costs (see.

---

[18]A certified green electricity tariff has been retrieved at www.verivox.de for area code 70173 (city of Stuttgart) and a consumption of 1000 kWh per year on April 01, 2015.

[19]Netze BW GmbH (2015), https://www.netze-bw.de/media/geteilte-medien/docs/veroeffentlichungspflichten/entgelte/preise-und-regelungen-gueltig-ab-01.01.2015.pdf.

[20]The calculations are shown in detail in Annex B—Cost Calculation.

Table 4.2 Operators of publicly accessible charging stations

Charging point operator	Charging points	Contract commitment	No contract commitment (calculated for BMW i3)	Household customer price	Mark-up (with contract)	Mark-up (without contract)
RWE AG	1,500	0.35 EUR/kWh	1.34 EUR/kWh	0.30 EUR/kWh	22 %	353 %
EnBW AG	750	0.36 EUR/kWh	1.35 EUR/kWh	0.30 EUR/kWh	22 %	358 %
Stromnetz Berlin GmbH (Vattenfall)	400	0.31 EUR/kWh	N.a.	0.30 EUR/kWh	5 %	N.a.
Stromnetz Hamburg GmbH (Hamburg Energie)	142	0.31 EUR/kWh	N.a.	0.30 EUR/kWh	5 %	N.a.
Stadtwerke Leipzig GmbH	136	N.a.	0.00 EUR/kWh	0.30 EUR/kWh	N.a.	−100 %
EWE AG	120	0.00 EUR/kWh	0.00 EUR/kWh	0.30 EUR/kWh	−100 %	−100 %
Stadtwerke Düsseldorf	100	N.a.	0.00 EUR/kWh	0.30 EUR/kWh	N.a.	−100 %
Stadtwerke München GmbH	52	N.a.	49 EUR onetime fee	0.30 EUR/kWh	N.a.	N.a.
RheinEnergie AG	46	N.a.	0.00 EUR/kWh	0.30 EUR/kWh	N.a.	−100 %
Mainova AG	45	N.a.	0.27 EUR/kWh	0.30 EUR/kWh	N.a.	−10 %
DREWAG Stadtwerke Dresden GmbH	44	N.a.	0.27 EUR/kWh	0.30 EUR/kWh	N.a.	−8 %
ENSO NETZ GmbH	36	N.a.	0.27 EUR/kWh	0.30 EUR/kWh	N.a.	−8 %
Aggregated number	3,371					
Last updated: 26.10.2015						

See Annex A—for sources

Table 4.3 Statement of costs for a charging tariff

Consumption-based costs		Netze BW—2015	Description
1.	Network tariff	5.58 Cent/kWh	Official price list
2.	Concession fee	2.39 Cent/kWh	See § 2 Abs. 2 Nr. 1 b) KAV
3.	Network fee for special usages	0.237 Cent/kWh	See § 19 Abs. 2 StromNEV
4.	CHP levy	0.254 Cent/kWh	See § 9 KWKG
5.	AbLaV levy	−0.051 Cent/kWh	See § 18 AbLaV
6.	Offshore levy	0.006 Cent/kWh	See § 17f EnWG
7.	EEG levy	6.17 Cent/kWh	See § 60 EEG
8.	Electricity tax	2.05 Cent/kWh	See§ 3 StromStG
9.	Procurement costs (wholesale)	3.96 Cent/kWh	Phelix future baseload + 0.5 Cent/kWh structuring costs
10.	Gross margin (incl. marketing)	3.90 Cent/kWh	According to FNA data
11.	Costs for the charging station	4.59 Cent/kWh	Own calculation
12.	Total	29.09 Cent/kWh	
Annual costs			
13.	Metering equipment	20.33 EUR/a	Smart meter (EDL21)
14.	Metering service	12.00 EUR/a	Monthly metering costs
15.	Billing	12.00 EUR/a	Monthly billing costs
16.	Consumption	1,000 kWh	
17.	Consumption-based costs	290.86 EUR	
18.	Annual costs	44.33 EUR	
19.	Total	335.19 EUR	
20.	VAT	63.69 EUR	
21.	Total incl. VAT	398.88 EUR	

The calculation of the cost of the hardware can be found in Annex B —cost calculation

line 20) are therefore around 400 EUR for a consumption of 1,000 kWh, hence at 0.40 cents/kWh.

Substantial Markup

In principle the FCJ and the ECJ require substantial markups which are an expression of antitrust disapproval. These mark-ups range from 5 to 25 %.[21] For the charging products with the contract commitment a determination of a substantial

[21]FCJ, Decision of December 07, 2010, case no. KZR 5/10, recital 33 (5 % markup); KG WuW/E OLG 2053, 2065/66 (25 %); Higher Regional Court Stuttgart, Decision of September 05, 2013, case no. 201 Kart 1/12, recital 9 (7.5 %); ECJ, Judgement of the Court of February 14, 1978, Cs.27/76 („United Brands"), which states a price can be deemd excessive "*because it has no reasonable relation to the economic value of the product*".

markup is no longer needed because it has been shown with reference to the price-cost analysis that prices in the amount of up to 0.40 cents/kWh are at a competitive level. Prices of charging products without a contract are, however, more than 300 % above the determined costs, so that a substantial price exaggeration is given—no matter which substantial markup was established.

Objective Justification

It is controversially discussed in the literature, whether there may be a possibility of justification for abusive conduct under Art. 102 TFEU,[22] while ECJ and FCJ as well as the European Commission explicitly support this possibility.[23] As a result, the possibility of a justification in regard to the abuse of dominance under Article 102 TFEU—either by way of a balancing of interests or a genuine justification - is generally agreed upon.[24]

Since in the present case the price exaggeration is visible only in regard to the charging products without a contract, the question arises whether the lack of contractual commitment provides somehow an objectively justifiable reason for the price mark-up.

Electricity products must be forecast as part of balancing group management, since the expected consumption of the customer must be projected as closely as possible. Deviations from forecast consumption are reflected in the form of so-called balancing energy costs. Without a contractual commitment the forecast of consumption could be quite challenging and therefore more prone to forecasting errors. But a glance at current balancing energy prices[25] shows that even a lack of forecasting—i.e. the total consumption of an electric vehicle is supplied solely with balancing energy—cannot remotely explain a double or even triple cost increase.

Such justifications are therefore not observable.

Interim Result

As pointed out in the present tariff comparison and price-cost analysis, there are some notable cases of excessive prices in the area of charging products without a

[22]Rejecting: Frenz (2005), recital 1414; consenting: Fuchs and Möschel (2012), recital 163.

[23]ECJ, Judgement of the Court of March 15, 2007, C 95/04 P, recital 69; FCJ, Judgement of December 07, 2010, case no. KZR 5/10 („Entega II"), recital 23; European Commission, Guidance on the Commission's enforcement priorities (2009/C 45/02), recital 28.

[24]Fuchs and Möschel (2012), recital 152; Wessely (2005), recital 15.

[25]E.g. Amprion GmbH (2015), http://www.amprion.net/ausgleichsenergiepreis.

contractual commitment. The excesses are based on costs, which would not occur in a competitive environment.

This is a price abuse under Art. 102 TFEU.

4.2.3.2 Price Discrimination Pursuant to Art. 102 TFEU

As shown in the previous section, there is a price differentiation between customers who choose a binding contract, and customers who choose no contractual commitment.

It is clear from the wording of Art. 102 para. 2 lit. c TFEU, that consumers are not within the scope of this variant of price differentiation, but this variant is exclusively for trade partners. But in addition to the already described abuse by charging excessive prices, pursuant to Article 102 para. 2 lit. a TFEU, the price differentiation may also be determined as an exclusionary conduct under Art. 102 para. 2 lit. b TFEU.

Exclusionary Conduct Against Charging Station Operators

The question is therefore whether the price discrimination between customers with and without a contract, constitutes exclusionary conduct via limitation of production, markets or technical development to the prejudice of consumers pursuant to Art. 102 para. 2 lit. b TFEU.

The contract commitment causes the client to pay monthly fees, regardless of whether in fact the charging stations are used or not. This product is therefore aimed at customers who can foresee regular use of the charging stations in a reasonably reliable way. These are customers that reside in the grid area of the charging infrastructure operator.

The charging products without a contract on the other hand are addressing customers who plan to charge less frequently or do not know at all when they will charge again. This will especially be customers from outside the respective grid area, thus customers who are in transit. For these customers a charging product with competitive kWh or minute prices, but with contract obligations, is less attractive because the monthly basic price would erode any price advantages. Customers traveling from outside the grid area who then want to charge their electric vehicles locally will therefore always choose a charging product without a contract.

However, this effect leads to market foreclosure in regard to the construction of charging stations by third parties in the grid area of the dominant charging station operator: As is clear from the price comparison, the charging stations operators offer their charging products with contract obligations at competitive prices or even lower, so that there is no incentive for a third party who wants to build charging stations in the grid area. Local customers are already contracted, and this at a price that the newcomer cannot undercut, at least not without generating losses.

Table 4.4 Price differentiation at charging stations

Charging point operator	Charging points	Price charging point operator (contract commitment)	Price charging point operator (no contract commitment)	Price via iPhone app of the new motion
RWE AG	1,500	0.36 EUR/kWh	1.34 EUR/kWh	0.56 EUR/kWh
EnBW AG	750	0.40 EUR/kWh	1.35 EUR/kWh	1.30 EUR/kWh
Stromnetz Berlin GmbH (Vattenfall)	400	0.38 EUR/kWh	N.a.	N.a
Stromnetz Hamburg GmbH (Hamburg Energie)	142	0.36 EUR/kWh	N.a.	N.a
Stadtwerke Leipzig GmbH	136	0	0.00 EUR/kWh	1.62 EUR/kWh
EWE AG	120	0.00 EUR/kWh	0.00 EUR/kWh	N.a
Stadtwerke Düsseldorf AG	100	N.a.	0.00 EUR/kWh	N.a.
Stadtwerke München GmbH	52	N.a.	49 EUR onetime fee	0.48 EUR/kWh
RheinEnergie AG	46	N.a.	0.00 EUR/kWh	N.a
Mainova AG	45	N.a.	0.27 EUR/kWh	N.a.
DREWAG—Stadtwerke Dresden GmbH	44	N.a.	0.27 EUR/kWh	N.a
ENSO NETZ GmbH	36	N.a.	0.27 EUR/kWh	N.a
Aggregate of charging points	3,371			
Last updated: 26.10.2015				

The price discrimination between local customers and customers from outside is thus leading to a lower expansion of charging stations by third parties.

Exclusionary Conduct Against Roaming Partners

The price differentiation described above is also found in the design of the roaming agreements again. The excessively high price has to be paid by the contracted roaming partners, i.e. other energy suppliers, who want to use the charging infrastructure in place to supply to their own customers. Third-party suppliers, such as the LichtBlick SE, for example, use the service of The New Motion to offer their own customers a charging possibility at public charging stations.[26]

[26]Zeitung für kommunale Wirtschaft (2015), https://www.zfk.de/nc/newsticker/artikel/lichtblick-oeffnet-seinen-kunden-15000-ladesaeulen.html?type=98.

The following analysis of the tariffs that are specified by The New Motion iPhone app,[27] partly shows significant price differences between the price offered directly by the charging point operator and the price offered via the iPhone App of The New Motion (Table 4.4).

These prices are billed to the third party suppliers, such as Lichtblick[28] or even The New Motion itself.[29] The third-party suppliers have to pass these prices on to the final consumer 1:1 if they don't want to make a loss. But a transaction fee has to be added as well, which is payable to the provider of the respective roaming platform. This amounts to about 30 to 50 euros per year and customer. Depending on what the third party supplier is able to pass on to its end customers, this pricing of currently available roaming solutions may lead to a more or less strong margin squeeze. The margin squeeze is applied firstly, through the pricing of charging stations providers themselves and on the other hand additionally by the transaction charges incurred for the use of the roaming platform.

This is in any case exclusionary conduct by the charging stations operators in respect of third suppliers in accordance with Art. 102 para. 2 lit. a TFEU.

References

Amprion GmbH (2015) Preise für Bilanzkreisabweichungen nach MaBiS ab 01.08.2011. http://www.amprion.net/ausgleichsenergiepreis. Accessed 07 July 2016

Bulst FW (2014) Art 102. In: Langen E, Bunte H-J (eds) Kartellrecht. Band 2. Europäisches Kartellrecht, 12th ed. Luchterhand, Köln

Eilmansberger T, Bien F (2015) Art. 102 AEUV. Münchener Kommentar. In: Bornkamm J, Montag F, Säcker FJ (eds) Europäisches und Deutsches Wettbewerbsrecht. Band 1: Europäisches Wettbewerbsrecht, 2nd ed. C.H.BECK, München

European Commission (2004) Guidelines on the assessment of horizontal mergers under the council regulation on the control of concentrations between undertakings. 2004/C 31/03. http://eur-lex.europa.eu/legal-content/EN/TXT/HTML/?uri=CELEX:52004XC0205%2802%29&from=DE. Accessed 07 July 2016

Frenz W (2005) Handbuch Europarecht. Band 2. Europäisches Kartellrecht. Springer, Berlin

Fuchs A, Möschel W (2012) Wettbewerbsrecht. Band 1 EU/Teil 1. In: Immenga U, Mestmäcker E-J (eds) Kommentar zum Europäischen Kartellrecht, 5th ed. C.H.BECK, München

Hamburger Abendblatt (2015) Laternen als Zapfsäule? Hamburg zögert. http://www.abendblatt.de/hamburg/article205337527/Laternen-als-Zapfsaeule-Hamburg-zoegert.html. Accessed 07 July 2015

LichtBlick SE (2015) LichtBlick-FahrStrom®—Die Energiewende wird mobil—100 % Tariftransparenz. http://www.lichtblick.de/privatkunden/strom/fahrstrom. Accessed 07 July 2016

Netze BW GmbH (2015) Preise und Regelungen für die Nutzung des Stromverteilnetzes der Netze BW GmbH. https://www.netze-bw.de/media/geteilte-medien/docs/veroeffent-lichungspflichten/entgelte/preise-und-regelungen-gueltig-ab-01.01.2015.pdf. Accessed 07 July 2016

[27]The price data was retrieved via the iPhone-App of The New Motion GmbH on August 06, 2015.

[28]LichtBlick (2015), http://www.lichtblick.de/privatkunden/strom/fahrstrom.

[29]The New Motion GmbH (2016), http://www.thenewmotion.com/de/produkte/ladekarte/.

Plugsurfing GmbH (2016) FAQ—Wie wird der Aufladepreis bestimmt? https://www.plugsurfing.com/de/faq. Accessed 07 July 2016

RWE Effizienz GmbH (2016) Ladesäulenfinder. https://www.rwe-mobility.com/web/cms/de/1195202/emobility/rwe-ladesaeulenfinder/. Accessed 07 July 2016

Schröter H, Bartl U (2015) In: von der Groeben H, Schwarze J, Hatje A (eds) Europäisches Unionsrecht. Band 2: Art. 67 bis 105 AEUV, 7th ed. Nomos, Baden-Baden

Smartlab Innovationsgesellschaft mbH (2015) Grenzenlos laden mit ladenetz.de. http://www.ladenetz.de/laden. Accessed 07 July 2016

Stadtwerke Aachen AG (2015) Antrag auf Nutzung der STAWAG Ladekarte für die Ladestationen. http://www.stawag-emobil.de/fileadmin/emobilitaet/content/Dokumente/Antrag_und_AGB_Ladekarte.pdf; Retrieved 08 June 2015

Technische Universität Berlin (2011) Die politischen Rahmenbedingungen der Genehmigung von Ladeinfrastruktur für Elektroverkehr im öffentlichen Raum. http://www.ivp.tu-berlin.de/fileadmin/fg93/Forschung/Projekte/e-mobility/Ergebnisbericht_Policy-Analyse__2_.pdf. Accessed 07 July 2016

Terhechte JP (2008) In: Terhechte JP (ed) Internationales Kartell-und Fusionskontrollverfahrensrecht. Gieseking, Bielefeld

The New Motion GmbH (2015) Fragen und Antworten—Welche öffentliche Ladestationen kann ich nutzen? http://www.thenewmotion.com/de/kontakt/fragen-und-antworten/. Accessed 07 July 2016

The New Motion GmbH (2016) Ladetarife. http://www.thenewmotion.de/pro-dukte/ladekarte/. Accessed 07 July 2016

Ubitricity Gesellschaft für verteilte Energiesysteme mbH (2015) Support—FAQs. https://ubitricity.com/de/support/faqs. Accessed 07 July 2016

Ubitricity Gesellschaft für verteilte Energiesysteme mbH (2016) Lösungen. https://ubitricity.com/de/l%C3%B6sungen. Accessed 07 July 2016

Wessely TW (2005) Art 82. Anwendungsgrundsätze. In: Jaeger W, Kokott J, Pohlmann P, Schroeder D (eds) Frankfurter Kommentar zum Kartellrecht, vol 57. Otto Schmidt, Köln

Zeitung für kommunale Wirtschaft (2015) Lichtblick öffnet seinen Kunden 15 000 Ladesäulen. https://www.zfk.de/nc/newsticker/artikel/lichtblick-oeffnet-seinen-kunden-15000-ladesaeulen.html?type=98. Accessed 07 July 2016

Chapter 5
Insufficient Expansion of Charging Stations

The assertion that publicly accessible charging stations are insufficiently expanded is frequently found in connection with the issue of e-mobility.[1] There are also numerous instances reported where public funds have been used to advance the expansion of the charging infrastructure.[2] It should therefore be examined whether a sufficient number of public charging stations is lacking, and if so, whether public intervention—and possibly in which form—is advised.

The claim to be examined reads as follows:

> ***An insufficient number of publicly accessible charging stations is installed and operational. This constitutes a market failure which can and should be remedied by government intervention.***

5.1 Benefit for the Final Consumer

The benefit for the final consumer would be that drivers of e-vehicles could rely on an extensive network of publicly accessible charging infrastructure in the future.

5.2 Relevant Market

Drivers of electric vehicles demand many opportunities to recharge their vehicles on the road, that is, in the middle of a journey or at the end of a trip. Looking at the range of one battery charge, a dense, nationwide network of public charging stations is needed, which should be comparable in density with the cover of conventional petrol stations. For the determination of the relevant market the findings in Sect. 3.2.1 can be applied here as well. Thereafter, the relevant product market is the construction

[1]faz.net (2015), Spiegel Online (2014), Sueddeutsche.de (2014).
[2]wiwo.de (2014), Hamburger Abendblatt (2014), cleanenergy-project.de (2015).

M. Adam, *Accelerating E-Mobility in Germany*, SpringerBriefs in Law
DOI 10.1007/978-3-319-44884-8_5

and operation of charging stations in the present case, in geographic terms the federal territory is likewise the relevant market.

5.3 Market Failure

After the first fundamental theorem of welfare economics, a competitive equilibrium leads, thus a perfect market, to a Pareto-optimal allocation.[3] Pareto-optimal in this context means that a property cannot be improved without deteriorating another property at the same time.[4] The Pareto optimum means in case of a market that available resources are used in an economically optimal way.[5] A market failure is therefore present when the market mechanism of supply and demand does not lead to economically desirable results, i.e. the factors of production are not used in such a way that they bring the maximum return for the national economy.[6] In the present case a disturbance of the balance of supply and demand could be detected in that an increase in demand is not served by the market, although the appropriate resources on the supply side as well as a willingness to pay on the demand side are present. To get an impression of the supply and demand of publicly available charging stations, it is important first to determine what the state of the current expansion of publicly accessible charging stations is, and then to examine in the next step to what extent this current state differs from their development goals and the expansion requirement.

5.3.1 Current Expansion of the Charging Station Network

The recent expansion of public charging points is shown in Table 5.1.

Graphically, this is represented as shown in Fig. 5.1.

Thus less charging points have been built for significantly more vehicles. It turns out that the ratio of vehicles per charging point has worsened—by a ratio of 2 vehicles per charging point to a ratio of almost 6 to 1 (see Fig. 5.2).

In other words, less charging points are available for more electric vehicles. The expansion of charging stations has thus been carried out more slowly than the registration of new electric vehicles, and in the period from January to July 2015, the expansion of charging stations was almost stagnant.[7]

---

[3]Arrow and Debreu (1954, p. 265).

[4]Arrow (1951, p. 508).

[5]Varian (2010, p. 311), Tirole (1998, p. 6).

[6]Bator (1958, pp. 351–352).

[7]BDEW states in its message from September 23, 2105, see footnote 14, that no decommissioning of charging stations took place, but this was merely a cleanup of double data entries.

Table 5.1 Expansion of charging points

Date	Registered E-vehicles	Charging points
Mid of 2012	5.960	2.821[a]
End of 2012	7.497	3.819[b]
Mid of 2013	10.401	4.386[c]
End of 2013	13.548	4.454[d]
Mid of 2014	24.000	4.800[e]
End of 2014	26.006	5.550[f]
Mid of 2015	31.631 [g]	5.571 [h]

[a]BDEW (2012), https://www.bdew.de/internet.nsf/id/DE_Ladeinfrastruktur

[b]BDEW (2013a), https://www.bdew.de/internet.nsf/id/bdew-erhebung-elektromobilitaet-anzahl-der-oeffentlichen-ladepunkte-seit-juli-2012-um-ueber-35-proz (Access for BDEW-members only)

[c]BDEW (2013b), https://www.bdew.de/internet.nsf/id/bdew-erhebung-elektromobilitaet-mittlerweile-fast-4-400-oeffentlich-zugaengliche-ladepunkte-in-deut

[d]BDEW (2014), https://www.bdew.de/internet.nsf/id/20140415-pi-branche-treibt-den-ausbau-von-ladesaeulen-voran-de?open&ccm=900010020010

[e]E&M Powernews (2015), http://www.energie-und-management.de/?id=84&no_cache=1&terminID=108441

[f]BDEW (2015a), https://www.bdew.de/internet.nsf/id/20150410-pi-ausbau-der-ladeinfrastruktur-fuer-elektroautos-in-deutschland-auf-gutem-weg-de

[g]Own calculation, based on data by BDEW and vehicle registration data by the Federal Motor Transport Authority (2015), http://www.kba.de/SharedDocs/Publikationen/DE/Statistik/Fahrzeuge/FZ/2015_monatlich/FZ8/fz8_201507_pdf.pdf?__blob=publicationFile&v=3

[h]BDEW (2015b), https://www.bdew.de/internet.nsf/id/bdew-erhebung-elektromobilitaet-de

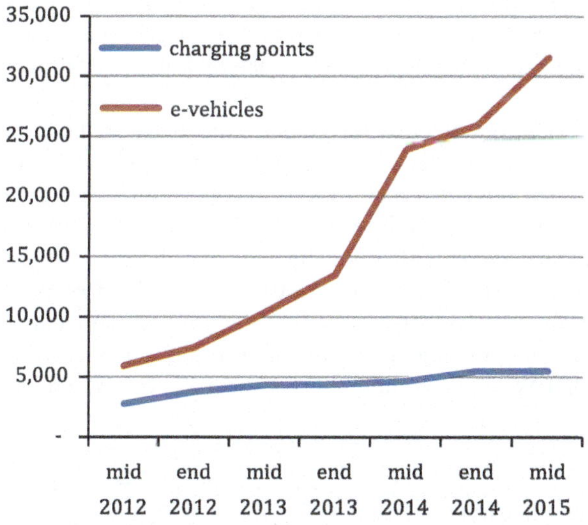

Fig. 5.1 Expansion of charging points and number of newly registered E-vehicles (*Source* Own graph based on data by BDEW)

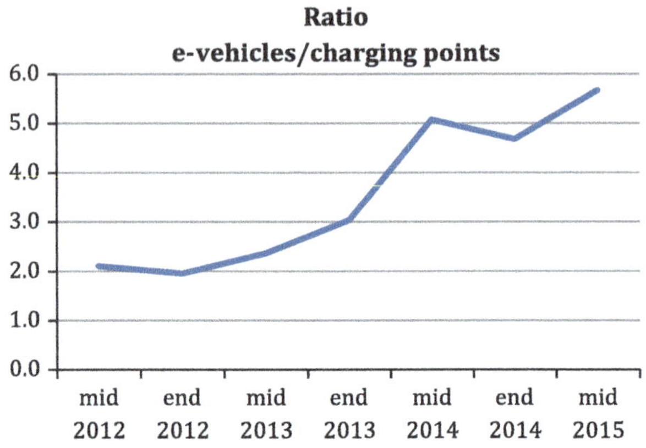

Fig. 5.2 Development of the ratio of E-vehicles to publicly accessible charging points (*Source* Own calculation and graph, based on data by BDEW)

Also with regard to the density of the charging station network it can be seen that the majority of the expansion is still to come: The e-mobility survey by the BDEW shows a concentration of charging points in major cities, but a lack of charging infrastructure in rural regions.[8] The white spots on the map make it clear that in many regions overland trips are not yet possible as charging stations between the metropolises are missing.

5.3.2 *Insufficient Expansion in Regard to the Current State*

With the development of the expansion of charging stations previously shown—the number of charging points is steadily increasing, but the expansion is slowing down appreciably—the question arises whether herein a market failure can already be seen.

The ratio of charging stations for vehicles has deteriorated from 1:2 to almost 1:6. The Commission considers a ratio of 1:10 as yet sufficient,[9] but a recent report from California just indicated that a ratio of 1:10 is not likely to be sufficient.[10]

The central question is therefore whether this slowing of expansion is due to a lower demand. This seems unlikely: The need for charging stations depends ultimately on the actual existing range of electric vehicles, i.e. for optimal mobility one or several charging stations must be within the range of an electric vehicle.[11] The lack of charging stations among the urban metropolises, clearly illustrated in the graph of the BDEW (see Footnote 11), is most likely not evidence of a lack of

[8]BDEW (2015c).

[9]European Commission (2014).

[10]The New York Times (2015).

[11]This consideration is also visible in the expansion plan for the Supercharger network, see Tesla Motors, Inc. (2013).

demand, but is rather an indication that the demand for charging stations for overland travel has not yet been met.

By comparison, vehicles with internal combustion engines have a range of 400 km or more before they need refueling. Currently, there are around 15,000 petrol stations in Germany,[12] and with approximately 6 fuel pumps per station this equates to roughly 90,000 refueling points. E-vehicles with a significantly lower range of roughly 100 to 200 km can currently only access about 5,500 publicly accessible charging points, of which the majority was built in cities and not on motorways or highways.

Since these vehicles shall gradually replace conventional vehicles with combustion engines, electric vehicles must therefore have at least the same features from a customer perspective. It must therefore be assumed that drivers of e-vehicles also want to drive long distances[13] and accordingly charging options en route are in demand.

But this demand is currently not being served.

5.3.3 *Insufficient Expansion in Regard to the Development Plan*

According to estimates by the European Commission 150,000 publicly accessible charging stations are required,[14] but according to lower estimates of the NPE there are still 70,000 charging stations required in the case scenario of 1 million electric vehicles.[15]

The discrepancy between the number of planned charging stations and charging stations built so far is enormous: The planned charging stations of at least 70,000 in 2020 compared to, as previously cited, only 5,550 charging points built by the end of 2014. With the current pace of development, the goal is not even close to being achieved.

It should be noted that the NPE, when determining the need for publicly accessible charging infrastructure, considered that only 15 % of the required charging infrastructure must be available to the public, as the majority of loads will be carried out in the garage at home or at work. These assumptions may be too optimistic:

---

[12]Mineralölwirtschaftsverband (2014, p. 37).

[13]The strategy of Tesla Motors can provide further evidence for such a demand: The Model S from Tesla Motors has been designed with a very large battery for a long range. Additionally the Supercharger network of Tesla Motors in Europe and North America supports traveling over long distances with many charging points.

[14]European Commission (2013).

[15]Federal Ministry for the Environment, Nature Conservation, Building and Nuclear Safety (2015, p. 45).

In 2013, 43 % of households were living in their own homes in Germany,[16] therefore less than half of households will also have their own garage. Even the charging at work is likely to constitute an exception in the long term: The consent of the property owner is needed for the installation of charging facilities and even if it is given, companies still have to actually carry out such an installation and in the next step the installed charging station must then be used by employees. The tax settlement of these charging processes at work (it is a free donation by the employer to the employee), however, is inefficient and represents a hurdle for both employers and employees.

All in all, the number of required charging points by the NPE is likely to be an underestimate.

5.3.4 Interim Result

The current state of development of charging stations is insufficient both in terms of the current coverage and in light of the expansion targets. Neither the basic need for mobility is currently covered, because driving between cities is not possible due to lack of charging stations in many regions, nor do the expansion targets for 2020 appear to be achievable.

The currently existing demand for charging points is not being adequately met and the necessary number of publicly accessible charging points in the future will not be achieved without any additional measures according to the current state.

5.4 Possible Reasons for the Lack of Expansion

In consideration of the points above, it follows that in the area of publicly accessible charging infrastructure a significant discrepancy between supply and demand must be observed. The market mechanism, which should lead to a Pareto-optimal allocation, is presently disturbed. Therefore it must be concluded that a market failure in the sense of the neoclassical theory of welfare economics exists. Hereinafter the possible causes of market failure are to be determined, so as to find an optimal solution for the market failure.

5.4.1 High Fixed Costs

Building and operating publicly accessible charging stations ultimately equates to the establishment of a network. And as is typical for networks and infrastructure in

[16]Statistisches Bundesamt (2014).

particular, the construction of charging stations is associated with high capital investment.[17] The establishment and operation of charging stations are connected to a high proportion of fixed costs, because initially at least 5,000 EUR need to be invested just for the construction of a single charging station, and of course further investments for the IT infrastructure are needed. These capital costs are fixed, that is, they are independent of the number of charging processes. With an increasing number of charging processes further costs for personnel and electricity procurement must be spent, but these costs are relatively small compared to the fixed costs.[18] If the charging stations are built once, the cost of operation compared to the cost of capital for the establishment is relatively insignificant. This accordingly leads to falling average costs, i.e. the more drivers are using the charging stations, the cheaper the deployment of the network per user or charging will be. If the average cost is constantly decreasing with increasing production volume, then economies of scale and thus subadditivity, i.e. a natural monopoly, independent of the level of production exist.[19]

As long as this monopoly market is contestable, that is, as long as there are potential competitors, who are willing to enter this market, exploitation of the monopoly situation is not to be expected.[20] In the present case, however, a competitor needs to make a high initial investment in order to participate in the market. This leads to the fact that competitors hold back with the entry into this market because, unlike the incumbent operators of the infrastructure newcomers must worry that they may lose all of the initial investment in the event of a market exit ("sunk costs").[21]

High fixed costs and sunk costs as a barrier to entry would immediately mean that new entrants would be kept away from the market. This phenomenon is currently visible since at present only energy utilities are building charging stations in public space to an appreciable extent, charging stations that are only installed in their own respective grid area, as for these companies the construction of charging stations is an extension of the electricity grid and the risk of sunk costs is therefore low: In awarding the electricity grid in accordance with § 46 EnWG the charging stations can be sold together with the grid as part of the distribution system.[22] However, this is in contrast with an operator of charging stations without an electricity distribution system: Such a company must fear that it could only sell its charging stations, at best, for a fraction of the value at market exit.

Such charging station operators without their own electricity distribution system are therefore also hardly observed on the German market.

---

[17]Prud'homme (2005, p. 4).

[18]Reinke (2014, p. 144).

[19]Diekmann et al. (2007, p. 18).

[20]Baumol (1982, p. 4).

[21]See Fritsch et al. (2007, p. 207), Tirole (1998, p. 308).

[22]This is what happened in fact in Hamburg, where the City of Hamburg acquired the charging stations from Vattenfall and transferred them to Stromnetz Hamburg GmbH, the new distribution system operator in the Hamburg.

High fixed costs combined with falling average costs and the irreversibility of the costs thus lead to the conclusion that an expansion of the charging infrastructure by new operators, without their own electricity distribution network, cannot be expected.

5.4.2 Economies of Scope

This effect of high fixed costs and irreversible costs is presently complemented by a further effect: The majority of operators of publicly available charging stations regularly benefit from synergy effects, i.e. the establishment of the charging station in the grid area of one's own group is less expensive than the establishment of a charging station in a grid area of another electricity distribution system operator in most cases. The synergies arise on the one hand when installing the charging stations, as the grid access planning can be done more easily and quickly inhouse than in the grid of another distribution system operator. Secondly, cost advantages can be gained in the operation, since the existing staff for the grid can manage the charging stations and the operation will run on the existing IT system. It is, to this extent, a geographical bundling effect.

This effect is also currently visible: The commitment of the charging stations operators focuses regularly (though not always exclusively) on its own distribution grid area or the distribution grid areas of the respective corporate group.

This negatively affects the expansion of the charging infrastructure as well, since the interest of a charging station operator to operate outside its own grid area, is low due to the lack of economies of scope.

5.4.3 Chicken-and-Egg-Problem

A similarly big problem as the two previously mentioned effects is the "chicken-and-egg-problem": While potential buyers of electric vehicles wait for adequate charging facilities until they actually buy an electric vehicle, the charging station operators are only inclined to build charging stations for the current, existing users. There is a vicious circle which is difficult to end and if at all, can be broken only after a long period of time.[23]

Since the construction of charging stations only generally pay off at a high utilization of the charging stations, the expansion is not progressing: The charging stations operators do not want to (or cannot) establish charging points in advance and except for the car buyers who purchase an electric vehicle out of conviction, the vast majority of buyers will still hold back.

---

[23]See Gillingham and Sweeney (2010, p. 76).

A nationwide network, which is difficult or may even be impossible to refinance by the current number of users, thus recedes into the distance.

This phenomenon is ultimately an expression of the situation of high fixed costs and concerns about sunk costs, with the difference in this case that there are no competitors facing each other, but suppliers and customers. Both suppliers and customers need to make a high initial investment to generally participate in this (new) market: The supplier needs to invest in a charging network with high fixed costs, the customer in a vehicle with high fixed costs - and both parties fear the probability of sunk costs.

So here it takes appropriate mechanisms to break this vicious circle.

5.4.4 Low Number of Available Public Places

As already described in Chap. 3, the parking space for publicly accessible charging points by the respective municipality is assigned exclusively to one provider in most cases. This practice also makes it generally hard for new suppliers to build a parallel network of charging infrastructure.

5.4.5 Interim Result

Each of the four phenomena described leads to the fact that entry of new suppliers into the market, here: charging stations operators, is hindered and due to the chicken-and-egg problem already active operators of charging infrastructure are also cautious with new investments.

This leads to the conclusion that new entrants on the market cannot be expected in significant magnitude, especially not by 2020.

Thus, further expansion of the charging infrastructure will only be carried out by the operators already active in the market, namely the energy utilities with their own electricity distribution system. So here the formation of monopolies or oligopolies becomes apparent and thus brings forth another form of market failure: The monopolist will not Pareto-optimally set the price at the level of marginal costs, but significantly above the marginal and average costs.[24] This profit-maximizing behavior results in a lower volume of supply. In other words: The monopolist will not provide the optimal amount, but achieve a higher monopoly price by a reduced amount. Thus, the formation of monopolies, in the case of publicly accessible charging infrastructure considered here, means that charging stations will not be built in sufficient quantity.

---

[24]Joskow (2007, p. 1252).

All in all, the existence of a classic market failure in terms of welfare economics is proven and the market failure can be attributed to several causes.

5.5 Appropriate Measures

It is now recognized that the existence of market failure alone is not sufficient to justify government intervention in market events. Rather, an additional assessment must be made as to whether the instrument chosen to eliminate or at least alleviate the market failure leads to less negative consequences than the market failure itself.[25] The possible tools to remedy the market failure should therefore be considered below and examined for their suitability.

5.5.1 Self-healing Powers of the Market

The principal idea is to question whether the self-healing powers of the market are not able to solve the various problems observed over time. Because of the ongoing formation of monopolies, however, waiting is not an option, because the self-healing powers can succeed here, if at all, only after a rather long period of time, for example, through new technical innovations.[26] However, because of the rather short period up to 2020, i.e. a period of just a few years, a reliance on the self-healing powers of the market—i.e. no government intervention—would be the wrong way in this case. On the contrary: A watching and waiting policy here would rather worsen the situation, as the reasons for the lack of expansion are varied and profound.

A reliance on the self-healing powers of the market therefore does not appear to be a viable solution.

5.5.2 Public Sponsorship

However, the task of establishing of a charging infrastructure is conceivable as a private endeavour, which could be funded to a certain extent by the public sector.

[25]Joskow (2007, p. 1249).

[26]An example of the self-healing powers could be the development of mobile phones and voice-over-IP technology as a technical alternative to traditional landline phones. A similar development could also occur in the current monopoly of the power grid as soon as batteries are offered in combination with solar cells at low costs. Although it will take some time for electricity consumers to actually get off the electrical grid and become energy self-sufficient.

The expansion of public charging infrastructure could thus be subsidized in part. The expansion would partly be financed with tax money.

5.5.3 Public Responsibility

The establishment and operation of publicly accessible charging infrastructure could also be perceived as a public task. Various embodiments would be conceivable in principle. The public task could for instance be fulfilled by a public company. The public task could also be performed by private companies, which could be assigned via a competitive tendering procedure.[27] In both cases, expansion and operation of charging stations would be fully financed with taxes or public dues.

5.5.4 Regulation

Between the two options of executing the tasks by private or public providers, there is a feasible middle course in the form of state regulation of this task.

In this case, the charging infrastructure could be regulated like the energy distribution systems, as described in Chap. 2. There would be regulated network charges, which would allow the charging station operator to generate reasonable returns for the network access of third party suppliers. There would be incentives for providing efficient network services via the established incentive regulation in the energy sector.[28] The abuse supervision by the regulators within the meaning of § 30 EnWG could counteract undesirable developments and furthermore there would be the possibility to integrate the expansion of the charging infrastructure into the network planning of the Federal Network Agency in accordance with § 12b EnWG. Thus, the charging infrastructure could be planned presciently.

5.5.5 Discussion

All options shown in Sects. 5.5.2–5.5.4 offer the opportunity to eliminate the described causes of the lack of expansion of the charging infrastructure, however, large differences show in terms of suitability.

---

[27]The City of Berlin chose this option, see Senatsverwaltung für Stadtentwicklung und Umwelt Berlin (2015).

[28]See ordinance on incentive regulation, AregV, and § 21 para. 2 EnWG.

Public support of the infrastructure development includes, in addition to the challenges of state aid law, the risk that the expansion is carried out in a very inefficient and costly way. Both the subsidization in part and the assignment of the infrastructure development as a governmental task entail the risk that the resources are not used optimally. State control of the efficient use of taxpayers' money is known to have its limits.

Carrying out the expansion of the charging infrastructure as a purely public task would incidentally constitute a break with the current approach. A constitutional justification for the transfer of this task from the current private enterprises to one or more public companies is difficult to imagine. The definition of the expansion of the charging infrastructure as a public task, however, appears theoretically feasible, but it is also missing a justification as to why the expansion should be a public responsibility, because a public good in the sense of welfare economics cannot be identified. Public goods are characterized by non-rivalry and non-exclusivity,[29] while the publicly accessible charging infrastructure rather shows the opposite characteristics: Consumers can be excluded, and there is a rival use instead. Since there is no public good, this speaks ultimately against the public tender of charging infrastructure services.

The regulation of the construction and operation of the charging infrastructure therefore presents itself as the most efficient solution: With the Energy Industry Act an existing and proven legal framework can be used and in addition to legal incentives for an efficient provision of services, a nation-wide (schematic) planning of the charging infrastructure expansion can also be carried out.

5.6 Result

Due to several forms of market failure, a state intervention in the development of publicly accessible charging infrastructure is warranted. The regulation of charging infrastructure provides the only course of action for the opportunity of a fast, efficient and demand-oriented expansion of the charging infrastructure. Other courses of action, such as tax-funded solutions or the tendering of charging infrastructure, however, are less optimal solutions.

References

Arrow KJ (1951) An extension of the basic theorems of classical welfare economics. In: Berkeley symposium on mathematical statistics and probability. Berkley, pp 507–532

Arrow KJ, Debreu G (1954) Existence of an equilibrium for a competitive economy. Econometrica. 22(3):265–290

---

[29]See Nicholson et al. (2009, p. 580), Fritsch et al. (2007, p. 365).

Bator FM (1958) The anatomy of market failure. Quart J Econ 72(3):351–379

Baumol W (1982) Contestable markets: an uprising in the theory of industry structure. Am Econ Rev 72:1–15

BDEW (2012) Zweite BDEW-Erhebung zur Ladeinfrastruktur für Elektromobilität. https://www.bdew.de/internet.nsf/id/DE_Ladeinfrastruktur. Accessed 07 July 2016

BDEW (2013a) Dritte BDEW-Erhebung zur Ladeinfrastruktur für Elektromobilität. https://www.bdew.de/internet.nsf/id/bdew-erhebung-elektromobilitaet-anzahl-der-oeffentlichen-ladepunkte-seit-juli-2012-um-ueber-35-proz. (Access for BDEW-members only). Accessed 20 May 2015

BDEW (2013b) BDEW-Erhebung Elektromobilität: Mittlerweile fast 4.400 öffentlich zugängliche Ladepunkte in Deutschland. https://www.bdew.de/internet.nsf/id/bdew-erhebung-elektromobilitaet-mittlerweile-fast-4–400-oeffentlich-zugaengliche-ladepunkte-in-deut. Accessed 07 July 2016

BDEW (2014) Branche treibt den Ausbau von Ladesäulen voran. https://www.bdew.de/internet.nsf/id/20140415-pi-branche-treibt-den-ausbau-von-ladesaeulen-voran-de?open&ccm=900010020010. Accessed 07 July 2016

BDEW (2015a) Ausbau der Ladeinfrastruktur für Elektroautos in Deutschland auf gutem Weg. https://www.bdew.de/internet.nsf/id/20150410-pi-ausbau-der-ladeinfrastruktur-fuer-elektroautos-in-deutschland-auf-gutem-weg-de. Accessed 07 July 2016

BDEW (2015b) Erhebung Elektromobilität. https://www.bdew.de/internet.nsf/id/bdew-erhebung-elektromobilitaet-de. Accessed 07 July 2016

BDEW (2015c) BDEW-Erhebung Elektromobilität: Ergebnisse und Grafiken (2. figure). http://www.voltimum.de/artikel/ausbau-der-ladeinfrastruktur-fuer-elektroautos-deutschland-auf-gutem-weg. Accessed 07 July 2016

cleanenergy-project.de (2015) EU fördert Ladestationen an deutschen Autobahnen. http://www.cleanenergy-project.de/mobilitaet/6611-eu-foerdert-ladestationen-an-deutschen-autobahnen. Accessed 07 July 2016

Diekmann J, Leprich U, Ziesing H-J (2007) Regulierung der Stromnetze in Deutschland. Düsseldorf: Hans-Böckler-Stiftung

European Commission (2013) EU launches clean fuel strategy. http://europa.eu/rapid/press-release_IP-13-40_en.htm. Accessed 07 July 2016

European Commission (2014) Clean fuels for transport: member states now obliged to ensure minimum coverage of refuelling points for EU-wide mobility. http://europa.eu/rapid/press-release_IP-14-1053_en.htm. Accessed 07 July 2016

E&M Powernews (2015) BDEW kritisiert Ladesäulenverordnung. http://www.energie-und-management.de/?id=84&no_cache=1&terminID=108441. Accessed 07 July 2016

Faz.net (2015) Kurzschluss in der E-Mobilität. http://www.faz.net/aktuell/rhein-main/rhein-main-kurzschluss-in-der-e-mobilitaet-13436868.html. Accessed 07 July 2016

Federal Ministry for the Environment, Nature Conservation, Building and Nuclear Safety (2015) Fortschrittsbericht 2014 –Bilanz der Marktvorbereitung, NPE. http://www.bmub.bund.de/themen/luft-laerm-verkehr/verkehr/luft-verkehr-download/artikel/fortschrittsbericht-2014-bilanz-der-marktvorbereitung/?tx_ttnews[backPid]=708. Accessed 07 July 2016

Federal Motor Transport Authority (2015) Neuzulassungen von Kraftfahrzeugen und Kraftfahrzeuganhängern - Monatsergebnisse Juli 2015. http://www.kba.de/SharedDocs/Publikationen/DE/Statistik/Fahrzeuge/FZ/2015_monatlich/FZ8/fz8_201507_pdf.pdf?__blob=publicationFile&v=3. Accessed 07 July 2016

Fritsch M, Wein T, Ewers H-J (2007) Marktversagen und Wirtschaftspolitik, 7th edn. Vahlen, München

Gillingham K, Sweeney J (2010) Market failure and the structure of externalities. In: Moselle B, Padilla AJ, Schmalensee R (eds) Harnessing renewable energy. Stanford, pp 69–92

Hamburger Abendblatt (2014) Hamburg investiert in Ausbau von Ladestationen für E-Autos. http://www.abendblatt.de/wirtschaft/article131632438/Hamburg-investiert-in-Ausbau-von-Ladestationen-fuer-E-Autos.html. Accessed 07 July 2016

Joskow PL (2007) Regulation of natural monopolies. In: Polinsky AM, Shavell S (eds) Handbook of law and economics (vol 2), pp 1227–1348

Mineralölwirtschaftsverband (2014) Jahresbericht 2013. http://www.mwv.de/upload/Publikationen/dateien/MWV_Jahresbericht_2013-Titelbild_mittel_5nWqRqG29p86sH5.pdf. Accessed 07 July 2016

Nicholson W, Snyder C, Luke P, Wood M (2009) Intermediate microeconomics. South-Western Cengage Learning, Andover

Prud'homme R (2005) Infrastructure and development. In: Bourguignon F, Pleskovic B (eds) Lessons of experience. Proceedings of the 2004 annual bank conference on development economics. The World Bank and Oxford. University Press, Washington, pp 153–181

Reinke J (2014) Bereitstellung öffentlicher Ladeinfrastruktur für Elektrofahrzeuge. Berlin

Senatsverwaltung für Stadtentwicklung und Umwelt Berlin (2015) Alliander gewinnt das erste deutsche Wettbewerbsverfahren für die Erweiterung und den Betrieb von Ladeinfrastruktur für Elektroautos. http://www.stadtentwicklung.berlin.de/ak-tuell/pressebox/archiv_volltext.shtml?arch_1501/nachricht5465.html. Accessed 07 July 2016

Spiegel Online (2014) Mehr Ladestationen für Elektroautos: Jetzt aber schnell. http://www.spiegel.de/auto/aktuell/schnellladesaeulen-fuer-elektroautos-ausbau-der-infrastruktur-a-962595.html. Accessed 07 July 2016

Statistisches Bundesamt (2014) Einkommens- und Verbrauchsstichprobe 2013. https://www.destatis.de/DE/ZahlenFakten/GesellschaftStaat/EinkommenKonsumLebensbedingungen/Wohnen/Aktuell_EVS.html. Accessed 07 July 2016

Sueddeutsche.de (2014) Völlig untauglich. http://www.sueddeutsche.de/muenchen/ausbau-der-elektromobilitaet-voellig-untauglich-1.2092316. Accessed 07 July 2016

Tesla Motors, Inc. (2013) Tesla motors presents supercharging at IAA and announces plans for European network expansion. http://www.teslamotors.com/blog/tesla-motors-presents-supercharging-iaa-and-announces-plans-european-network-ex. Accessed 07 July 2016

The New York Times (2015) In California, electric cars outpace plugs, and sparks fly. http://www.nytimes.com/2015/10/11/science/in-california-electric-cars-outpace-plugs-and-sparks-fly.html?smid=tw-nytimes&smtyp=cur&_r=0. Accessed 07 July 2016

Tirole J (1998) The theory of industrial organization, 10th edn. MIT Press, Cambridge

Varian HR (2010) Intermediate microeconomics, 8th edn. Norton & Company, New York

wiwo.de (2014) Ladestationen zwischen München und Berlin. http://www.wiwo.de/unternehmen/auto/elektroautos-ladestationen-zwischen-muenchen-und-berlin/9895346.html. Accessed 07 July 2016

Chapter 6
Summary and Proposals

Aim of the study was the identification of possible negative developments in the field of publicly accessible charging infrastructure. The results of the study are summarized again below. In the second part of this section, conclusions for the further development of the legal framework for public charging infrastructure are drawn from the results.

6.1 Summary

The study has shown that the—just emerging—markets around the publicly accessible charging infrastructure are characterized by a lack of competition and the formation of a nationwide oligopoly. The market for the establishment and provision of the charging infrastructure is dominated by an oligopoly, the operators of the charging infrastructure. This oligopoly has eliminated competition in the market of supplying final customers by denying network access to third parties and by means of roaming agreements. Pricing is not executed through a competitive market; on the contrary every charging infrastructure operator is able to set a price itself. As shown in the study, this leads firstly to cases of highly excessive pricing and on the other hand to price differentiation, aiming to defend market positions of the established charging station operators against potential competitors.

As pointed out in the study, the oligopoly also affects more or less directly the further expansion of the charging infrastructure. The foreclosure of retail supply in combination with the high investment costs for setting up a supplier's own network of charging infrastructure leads to hesitant or no investments in new charging points at all.

This effect is reinforced by the "chicken-and-egg-problem", i.e. the hitherto low registrations of electric vehicles represent a significant utilization risk for charging infrastructure operators. Depending on the location, charging infrastructure operators may face the situation that an investment in a charging point will only be

M. Adam, *Accelerating E-Mobility in Germany*, SpringerBriefs in Law
DOI 10.1007/978-3-319-44884-8_6

profitable after many years, because sufficient revenues cannot be generated with the charging station since e-vehicles have not been widely adopted and new registrations of electric vehicles still remain low.

The classification of public charging stations as electricity distribution systems within the meaning of the EnWG and the Directive 2009/72/EC addresses all the problems described herein and promises the most efficient solution to these problems. While competition law with the supervision of abusive practices provides an instrument which can help here in principle, the efficient implementation will be difficult given the large number of charging station operators. As a look at the liberalization of the electricity and gas supply in the years 1998 to 2005 shows, regulation is the better choice in the long run.

Last but not least energy regulation is the only way to use key advantages of E-Mobility at all, because the grid integration of electric vehicles and also the 100 % supply of green electricity can only be achieved by means of regulated network access.

The liberalization of the electricity supply in Germany was to be ensured via the negotiated network access based on the association agreements in the years 1998–2005. This form of self-regulation, which was nothing more than a lack of regulation, failed on a large scale and yet looking at the publicly accessible charging infrastructure, this story repeats itself, only that the association agreements are now called roaming agreements and instead of negotiated network access, the single buyer model (or resale model) is immediately relied on.

In light of the mistakes of the past, politics should not delay the regulation, but take action immediately. To say it in the own words of the Federal Government:

> *„In particular, future public charging stations should be non-discriminatory accessible for each electricity supplier and each vehicle model. It would not be desirable if every supplier created a separate infrastructure of charging stations for its customers. The current supply of electric vehicles must also take place within competition just as energy supply to households does."*[1]

This goal, formulated in 2009, must now be turned into reality. A publicly accessible charging infrastructure with

- easy access with just one charge card,
- customer's free choice of electricity suppliers,
- reasonable and transparent prices,
- and sufficient charging points

can only be achieved through the regulation of charging infrastructure.

[1]Federal Ministry for Economic Affairs and Energy (2009, p. 40).

6.2 Proposals

The energy regulation of publicly accessible charging infrastructure is indeed a discussed and viable solution, however, so far there have been except for the pilot project in Hamburg no efforts to implement this system. Clarifications in the law can address uncertainties in the application of the law and thus set the necessary legal framework for the successful development of e-mobility in Germany.

6.2.1 Legal Classification of the Charging Station

As shown in Sect. 2.3, publicly accessible charging stations qualify as an energy distribution system for general supply. This classification could for example be carried out by adding the publicly accessible charging point expressly to the legal definition in § 3 no. 3 EnWG.

6.2.2 Charging Station as an Allowable Asset Pursuant to the StromNEV

Corresponding to the fundamental clarification, the charging station should be listed as an allowable asset in Appendix 1 to StromNEV under Paragraph III ("transmission and distribution systems"), so as to ensure that the charging stations can be part of the network tariff cost calculation in the future. Accompanying measures of the legislature should be the consideration of the charging station on the enlargement factor in accordance with § 19 ARegV in the Incentive Regulation or alternatively via the investment measure in accordance with § 23 ARegV.

Ultimately, the financing of public charging infrastructure through network tariffs must be made possible.

Particularly in view of the practical relevance of DC fast charging with high power charging it should be considered if research and development costs for fast charging stations could be more incentivized for a limited period, e.g. through an explicit allowance of these costs by a higher surcharge than the current 50 % pursuant to § 25a para. 1 sentence 2 ARegV.

6.2.3 Structure of the Electricity Distribution Systems

As mentioned above, Germany has an extremely high number of small and medium-sized distribution systems. This structure generates high transaction costs on nationwide journeys because with every change of the electricity distribution system, i.e. every city, metering and billing fees currently apply anew. It must also be questioned whether the many small municipal utilities will be able to perform numerous intraday changes of supplier.

In order to create greater operating units in the form of partnerships and other associations ("grid cluster"), the de minimis provision in § 7 para. 2 EnWG and § 7a Abs. 7 should therefore be abolished.

6.2.4 Abolishing the Billing Fee

So far, a billing fee has been charged for each settlement of network charges. This goes back to the rules in Appendix 2 of StromNEV, which states that a main cost centre "*settlement*" must be created for the "*cost of the commercial processing of meter data; cost of providing due fees for network use and billing*" in the network tariff calculation. This billing fee is very high and it additionally has an equivalent effect to a customs duty, if an electric vehicle travels from one network to another network and recharges in both grid areas, because sums of around 10 EUR/a are payable solely for the billing fee in each grid area.[2] With more than 880 distribution grids in Germany a long-distance trip is therefore very expensive.

As is being considered in the Key Issues Paper "Regulation Package Intelligent Networks" of the Federal Ministry for Economic Affairs and Energy,[3] the billing fee should be abolished and these costs should be reflected in the general network costs.

6.3 Closing Remarks

The aforementioned proposals can be implemented into the legislative process without great effort, since they are merely additions and clarifications of existing rules. These proposals may serve to improve the structure and use of publicly accessible charging infrastructure, but it should be noted that according to the results of this study regulating the publicly accessible charging infrastructure is necessary and already possible within the existing legal regulations.

An effective coupling of the transport and energy sectors, with the aim of creating a sustainable and renewable transport sector will only succeed if the driver of an electric vehicle is understood as an energy customer and charging infrastructure operators are understood as energy utilities. Accordingly the regulatory

[2]For example, Westnetz GmbH charges a billing fee of 12.83 EUR per year (incl. VAT), see List of Fees 2015, https://www.westnetz.de/web/cms/mediablob/de/2290308/data/1625968/9/westnetz/netz-strom/netzentgelte/preisblaetter-2015/Netznutzungspreise-gueltig-vom-01.01.15-bis-31.12.15-.pdf.

[3]Federal Ministry of Economic Affairs and Energy (2015, p. 4).

requirements of the Energy Industry Act and the Directive 2009/72/EC must be applied to the publicly accessible charging infrastructure.

The breakthrough of electric mobility will largely depend on this.

References

Federal Ministry of Economic Affairs and Energy (2015) Baustein für die Energiewende: 7 Eckpunkte für das "Verordnungspaket Intelligente Netze". http://www.bmwi.de/BMWi/Redaktion/PDF/E/eckpunkte-fuer-das-verordnungspaket-intelligente-netze,property=pdf,bereich=bmwi2012,sprache=de,rwb=true.pdf. Accessed 07 July 2016

Federal Ministry for Education and Research (2009) Nationaler Entwicklungsplan Elektromobilität der Bundesregierung. https://www.bmbf.de/files/nationaler_entwicklungsplan_elektromobilitaet.pdf. Accessed 07 July 2016

Westnetz GmbH (2015) List of fees (2015) https://www.westnetz.de/web/cms/mediablob/de/2290308/data/1625968/9/westnetz/netz-strom/netzentgelte/preisblaetter-2015/Netznutzungspreise-gueltig-vom-01.01.15-bis-31.12.15-.pdf. Accessed 07 July 2016

Annex A—Overview Operators of Charging Stations

Hereinafter, the largest charging infrastructure operators are listed in Germany. The list is not exhaustive:

Table A.1 Overview operators of charging stations

Charging station operator	Public	Tariff with	Tariff without	
	Charging points	Contract[a]	Contract[b]	Description
RWE AG[1]	1,500	0.36 Ct/kWh	1.34 Ct/kWh	Operated by RWE Effizienz GmbH
EnBW AG[2]	750	0.36 Ct/kWh	1.35 Ct/kWh	4th largest utility in Germany
Stromnetz Berlin GmbH[3]	400	0.31 Ct/kWh	n.a.	Grid operator Berlin (Vattenfall)
Stromnetz Hamburg GmbH[4]	142	0.31 Ct/kWh	n.a.	Grid operator in Hamburg
Stadtwerke Leipzig GmbH[5]	136	n.a.	0.00 Ct/kWh	Utility in Leipzig
EWE AG[6]	120	n.a.	0.00 Ct/kWh	5th largest utility in Germany
Stadtwerke Düsseldorf GmbH[7]	100	n.a.	0.00 Ct/kWh	Utility in Düsseldorf
Stadtwerke München GmbH[8]	52	n.a.	49 EUR	Utility in Munich
RheinEnergie AG[9]	46	n.a.	0.00 Ct/kWh	Utility in Cologne
Mainova AG[10]	45	n.a.	0.27 Ct/kWh	Grid operator in Frankfurt
DREWAG Stadtwerke Dresden GmbH[11]	44	n.a.	0.27 Ct/kWh	Utility in Dresden
ENSO NETZ GmbH[11]	36	n.a.	0.27 Ct/kWh	Grid operator in Saxonia

(continued)

M. Adam, *Accelerating E-Mobility in Germany*, SpringerBriefs in Law
DOI 10.1007/978-3-319-44884-8

Table A.1 (continued)

Charging station operator	Public	Tariff with	Tariff without	
	Charging points	Contract[a]	Contract[b]	Description
Total number charging points	3,371			Date: 26.10.2015

[a]The price calculation is based on an annual consumption of 1,000 kWh

[b]Hourly prices (Type-2-connector) have been converted with a capacity of 3.7 kW per hour (based on a BMW i3)

[1]RWE Effizienz GmbH, RWE bietet intelligente Ladelösungen: Übersicht Infrastruktur, p. 3; http://www.rwe.com/web/cms/mediablob/de/2102054/data/249840/6/rwe-effizienz-gmbh/presse-news/blob.pdf; retrieved October 26, 2015; RWE Effizienz GmbH, RWE epower basic; https://www.rwe-mobility.com/web/cms/de/1232368/pro-dukte-services/rwe-epower/rwe-epower-basic/; abgerufen am 26.10.2015; RWE Effizienz GmbH, RWE epower direct; https://www.rwe-mobility.com/web/cms/de/2682124/produkte-services/rwe-epower/rwe-epower-direct/; retrieved October 26, 2015.

[2]BEM Bundesverband eMobilität e.V., 100.000 Ladevorgänge an EnBW Ladestationen; http://www.bem-ev.de/100-000-ladevorgange-an-enbw-ladestationen-192/; retrieved October 26, 2015; EnBW Elektronauten-Ladekarte, Typ 2-Stecker bis 5 kW; https://www.enbw.com/privatkunden/tarife-und-produkte/e-mobilitaet/enbw-elektronauten-ladekarte/index.html; retrieved October 26, 2015.

[3]Stromnetz Berlin GmbH, 2015—Der Netzausbau geht weiter, p. 2; http://www.stromnetz-berlin.de/de/file/Jahrespresse-konferenz-2015-Praesentation_70254274.pdf; retrieved October 26, 2015; Vattenfall Europe Innovation GmbH, Ladekarte Berlin; http://www.vattenfall.de/de/emobility/ladekarte-bestellen.htm; retrieved October 26, 2015.

[4]Stromnetz Hamburg GmbH, Faktenblatt Elektromobilität; http://www.stromnetz.hamburg/ueber-uns/innovationen/e-mobility/; retrieved October 26, 2015; Hamburg Energie GmbH, Horizont Mobil—Auftragsformular; http://www.hamburgenergie.de/privatkunden/oekostrom/mobilitaets-tarif/; retrieved October 26, 2015.

[5]Stadtwerke Leipzig GmbH, Liste der Elektro-Ladestationen; https://www.swl.de/MediaLibrary/Content/web/swl/DE/Unternehmen/Elektromobilitaet/ladestationen/liste_ladestationen.pdf; retrieved October 26, 2015; Stadtwerke Leipzig GmbH, Ladestationen; https://www.swl.de/web/swl/DE/Unternehmen/Elektromobilitaet/lade-stationen/Ladestationen.htm; retrieved October 26, 2015.

[6]EWE AG, EWE Stromtankstellen; http://www.ewe.de/privatkunden/erdgas/dienste/mobilitaet/stromtankstellen; retrieved October 26, 2015.

[7]BEM Bundesverband eMobilität e.V., 100.000 Ladevorgänge an EnBW Ladestationen; http://www.bem-ev.de/100-000-ladevorgange-an-enbw-ladestationen-192/; retrieved October 26, 2015; Stadtwerke Düsseldorf AG, Antrag für die Tankkarte der Stadtwerke Düsseldorf AG; https://www.swd-ag.de/formulare/tankkarte/; retrieved October 26, 2015.

[8]Stadtwerke München GmbH, SWM Ladestationen für Elektroautos; http://www.swm.de/privatkunden/unternehmen/inno-vation/elektro-mobilitaet/swm-ladestationen.html; retrieved October 26, 2015; Stadtwerke München, Strom tanken; https://www.swm.de/privatkunden/unternehmen/innovation/elektromobilitaet/strom-tanken.html; retrieved October 26, 2015.

[9]Portal-21 Onlineverlag GmbH; Ford/RheinEnergie—CognE-mobil zieht Bilanz zu Elektromobilität im Kölner Raum; http://www.portal-21.de/elektromobilitaet/artikel/cologne-mobil-zieht-bilanz-zu-elektromobilitaet-im-koelner-raum-0119105812/; retrieved October 26, 2015; RheinEnergie AG, Elektromobilität, Registrierung "TankE"; http://www.rheinenergie.com/de/unternehmensportal/technik_zukunft/elektromobilitaet_1/registrierung_2/index.php; retrieved October 26, 2015.

[10]GoingElectric Stromtankstellenverzeichnis, Mainova Stromtankstellen in Deutschland; http://www.goingelectric.de/strom-tankstellen/Deutschland/Verbund/Mainova/; retrieved October 26, 2015; Mainova AG, E-Mobilität in Frankfurt; https://www.mainova.de/static/de-mainova/downloads/2011-02-11_Frankfurter_Modell.pdf; retrieved October 26, 2015.

[11]DREWAG, Elektromobilität—Standorte in Sachsen; https://www.drewag.de/de/drewag/neue-energien/elektromobilitaet-standorte.php; retrieved October 26, 2015; DREWAG, StromTicket; https://www.drewag.de/de/drewag/neue-energien/elektromobilitaet-strom-ticket.php; retrieved October 26, 2015.

Annex B—Cost Calculation

Table B.1 Cost calculation charging infrastructure and smart meter (EDL21)

Annual costs of a charging station (3.7 kW, 2 charging points)						
Acquisition and installation	6,500.00 EUR	Source: Masterplan E-Mobilität Hamburg, S. 21; http://www.hamburg.de/contentblob/4362700/data/pm-26-08-2014-masterplan.pdf				
Operating life	25	(see StromNEV Appendix 5 No. 2.3)				
WACC	6.00 %					
Process of depreciation	Year 1	Year 2	Year 3	Year 4	Year 5	Year 6
Residual value	6,240.00 EUR	5,980.00 EUR	5,720.00 EUR	5,460.00 EUR	5,200.00 EUR	4,940.00 EUR
Interest	374.40 EUR	358.80 EUR	343.20 EUR	327.60 EUR	312.00 EUR	296.40 EUR
Depreciation	260.00 EUR	260.00 EUR	260.00 EUR	260.00 EUR	260.00 EUR	260.00 EUR
Capital costs	634.40 EUR	618.80 EUR	603.20 EUR	587.60 EUR	572.00 EUR	556.40 EUR
Operating costs	100.00 EUR	100.00 EUR	100.00 EUR	100.00 EUR	100.00 EUR	100.00 EUR
Total	734.40 EUR	718.80 EUR	703.20 EUR	687.60 EUR	672.00 EUR	656.40 EUR
Costs per charging point	367.20 EUR	359.40 EUR	351.60 EUR	343.80 EUR	336.00 EUR	328.20 EUR
Consumption per charging point	8,000 kWh	8,000 kWh	8,000 kWh	8,000 kWh	8,000 kWh	8,000 kWh
Cost per kWh	4.59 Ct/kWh	4.49 Ct/kWh	4.40 Ct/kWh	4.30 Ct/kWh	4.20 Ct/kWh	4.10 Ct/kWh
Metering costs (EDL 21)						
Acquisition	140.00 EUR	(Average price based on a Google-research)				

(continued)

M. Adam, *Accelerating E-Mobility in Germany*, SpringerBriefs in Law
DOI 10.1007/978-3-319-44884-8

Table B.1 (continued)

Metering costs (EDL 21)						
Installation	50.00 EUR	(Incl. 1/2 h for installation, tooling and call-out-fee)				
Total	190.00 EUR					
Operating life	20	(See StromNEV Appendix 5 No. 2.6)				
WACC	6.00 %					
Process of depreciation	Year 1	Year 2	Year 3	Year 4	Year 5	Year 6
Residual value	180.50 EUR	171.00 EUR	161.50 EUR	152.00 EUR	142.50 EUR	133.00 EUR
Interest	10.83 EUR	10.26 EUR	9.69 EUR	9.12 EUR	8.55 EUR	7.98 EUR
Depreciation	9.50 EUR	9.50 EUR	9.50 EUR	9.50 EUR	9.50 EUR	9.50 EUR
Total	20.33 EUR	19.76 EUR	19.19 EUR	18.62 EUR	18.05 EUR	17.48 EUR